POWER QUALITY

The ELECTRIC POWER ENGINEERING Series
Series Editor Leo Grigsby

Published Titles

Electromechanical Systems, Electric Machines, and Applied Mechatronics
Sergey E. Lyshevski

Electrical Energy Systems
Mohamed E. El-Hawary

Electric Drives
Ion Boldea and Syed Nasar

Distribution System Modeling and Analysis
William H. Kersting

Linear Synchronous Motors: Transportation and Automation Systems
Jacek Gieras and Jerry Piech

The Induction Machine Handbook
Ion Boldea and Syed Nasar

Power Quality
C. Sankaran

Forthcoming Titles

Power System Operations and Electricity Markets
Fred I. Denny and David E. Dismukes

Computational Methods for Electric Power Systems
Mariesa Crow

POWER QUALITY

C. SANKARAN

CRC Press
Taylor & Francis Group
Boca Raton London New York

CRC Press is an imprint of the
Taylor & Francis Group, an **informa** business

CRC Press
Taylor & Francis Group
6000 Broken Sound Parkway NW, Suite 300
Boca Raton, FL 33487-2742

© 2002 by Taylor & Francis Group, LLC

CRC Press is an imprint of Taylor & Francis Group, an Informa business

No claim to original U.S. Government works

ISBN-13: 978-0-8493-1040-9 (hbk)
ISBN-13: 978-0-367-39646-6 (pbk)

Visit the Taylor & Francis Web site at
http://www.taylorandfrancis.com

and the CRC Press Web site at
http://www.crcpress.com

Library of Congress Card Number 2001043744

Library of Congress Cataloging-in-Publication Data

Sankaran, C.
　Power quality / C. Sankaran.
　　p. cm.
　Includes index.
　ISBN 0-8493-1040-7 (alk. paper)
　1. Electric power system stability. 2. Electric power systems—Quality control. I. Title.

TK1010 .S35 2001
621.31'042—dc21　　　　　　　　　　　　　　　　　　　　　　2001043744

Dedication

This book is dedicated to
God, who is my teacher;
Mary, my inspiration and friend; and
Bryant, Shawn, and Kial, who are everything that a father could want.

Preface

The name of this book is *Power Quality,* but the title could very well be *The Power Quality Do-It-Yourself Book.* When I set out to write this book, I wanted it to be user friendly, easy to understand, and easy to apply in solving electrical power system problems that engineers and technicians confront on a daily basis. As an electrical engineer dealing with power system quality concerns, many of the books I consulted lacked direct and precise information and required a very thorough search to find what I needed. Very often, I would spend several hours pondering a case just so the theory I read and the practical findings would come together and make sense. This book is the product of my thought processes over many years. I have tried to combine the theory behind power quality with actual power quality cases which I have been involved with in order to create a book that I believe will be very useful and demystify the term *power quality.*

What is power quality? Power quality, as defined in this book, is "a set of electrical boundaries that allows equipment to function in its intended manner without significant loss of performance or life expectancy." Conditions that provide satisfactory performance at the expense of life expectancy or vice versa are not acceptable.

Why should power quality be a concern to facility designers, operators, and occupants? When the quality of electrical power supplied to equipment is deficient, performance degradation results. This is true no matter if the equipment is a computer in a business environment, an ultrasonic imaging machine in a hospital, or a process controller in a manufacturing plant. Also, good power quality for one piece of equipment may be unacceptable for another piece of equipment sitting right next to it and operating from the same power lines, and two identical pieces of equipment can react differently to the same power quality due to production or component tolerances. Some machines even create their own power quality problems. Given such hostile conditions, it is important for an engineer entrusted with the design or operation of an office building, hospital, or a manufacturing plant to be knowledgeable about the basics of power quality.

This book is based on 30 years of personal experience in designing, testing, and troubleshooting electrical power systems and components, the last 9 of which have been spent exclusively studying and solving power quality problems for a wide spectrum of power users. This book is not an assemblage of unexplained equations and statements. The majority of the information contained here is based on my experiences in the power system and power quality fields. Mathematical expressions are used where needed because these are essential to explaining power quality and its effects. Throughout the book, several case examples are provided, the steps used to solve power quality problems are described in depth, and photographs, illustrations, and graphs are used to explain the various power quality issues. The examples show that many power quality problems that have resulted in loss of productivity, loss of equipment, injury to personnel, and in some cases, loss of life could easily have been avoided. All that is needed to prevent such consequences is a clear understanding of electrical power quality and its effects on power system performance.

I hope the reader will enjoy reading this book as much as I enjoyed writing it. Also, I hope the reader will find the book useful, as it is based on the experiences of an electrical engineer who has walked through the minefields of electrical power system quality and for the most part survived.

C. Sankaran

Contents

1 Introduction to Power Quality

1.1 DEFINITION OF POWER QUALITY

Power quality is a term that means different things to different people. Institute of Electrical and Electronic Engineers (IEEE) Standard IEEE1100 defines power quality as "the concept of powering and grounding sensitive electronic equipment in a manner suitable for the equipment." As appropriate as this description might seem, the limitation of power quality to "sensitive electronic equipment" might be subject to disagreement. Electrical equipment susceptible to power quality or more appropriately to lack of power quality would fall within a seemingly boundless domain. All electrical devices are prone to failure or malfunction when exposed to one or more power quality problems. The electrical device might be an electric motor, a transformer, a generator, a computer, a printer, communication equipment, or a household appliance. All of these devices and others react adversely to power quality issues, depending on the severity of problems.

A simpler and perhaps more concise definition might state: "Power quality is a set of electrical boundaries that allows a piece of equipment to function in its intended manner without significant loss of performance or life expectancy." This definition embraces two things that we demand from an electrical device: performance and life expectancy. Any power-related problem that compromises either attribute is a power quality concern. In light of this definition of power quality, this chapter provides an introduction to the more common power quality terms. Along with definitions of the terms, explanations are included in parentheses where necessary. This chapter also attempts to explain how power quality factors interact in an electrical system.

1.2 POWER QUALITY PROGRESSION

Why is power quality a concern, and when did the concern begin? Since the discovery of electricity 400 years ago, the generation, distribution, and use of electricity have steadily evolved. New and innovative means to generate and use electricity fueled the industrial revolution, and since then scientists, engineers, and hobbyists have contributed to its continuing evolution. In the beginning, electrical machines and devices were crude at best but nonetheless very utilitarian. They consumed large amounts of electricity and performed quite well. The machines were conservatively designed with cost concerns only secondary to performance considerations. They were probably susceptible to whatever power quality anomalies existed at the time, but the effects were not readily discernible, due partly to the robustness of the

1

machines and partly to the lack of effective ways to measure power quality parameters. However, in the last 50 years or so, the industrial age led to the need for
products to be economically competitive, which meant that electrical machines were
becoming smaller and more efficient and were designed without performance mar-
gins. At the same time, other factors were coming into play. Increased demands for
electricity created extensive power generation and distribution grids. Industries
demanded larger and larger shares of the generated power, which, along with the
growing use of electricity in the residential sector, stretched electricity generation
to the limit. Today, electrical utilities are no longer independently operated entities;
they are part of a large network of utilities tied together in a complex grid. The
combination of these factors has created electrical systems requiring power quality.

The difficulty in quantifying power quality concerns is explained by the nature
of the interaction between power quality and susceptible equipment. What is "good"
power for one piece of equipment could be "bad" power for another one. Two
identical devices or pieces of equipment might react differently to the same power
quality parameters due to differences in their manufacturing or component tolerance.
Electrical devices are becoming smaller and more sensitive to power quality aberrations due to the proliferation of electronics. For example, an electronic controller
about the size of a shoebox can efficiently control the performance of a 1000-hp
motor; while the motor might be somewhat immune to power quality problems, the
controller is not. The net effect is that we have a motor system that is very sensitive
to power quality. Another factor that makes power quality issues difficult to grasp
is that in some instances electrical equipment causes its own power quality problems.
Such a problem might not be apparent at the manufacturing plant; however, once
the equipment is installed in an unfriendly electrical environment the problem could
surface and performance suffers. Given the nature of the electrical operating boundaries and the need for electrical equipment to perform satisfactorily in such an
environment, it is increasingly necessary for engineers, technicians, and facility
operators to become familiar with power quality issues. It is hoped that this book
will help in this direction.

1.3 POWER QUALITY TERMINOLOGY

Webster's New World Dictionary defines *terminology* as the "the terms used in a
specific science, art, etc." Understanding the terms used in any branch of science or
humanities is basic to developing a sense of familiarity with the subject matter. The
science of power quality is no exception. More commonly used power quality terms
are defined and explained below:

> **Bonding** — Intentional electrical-interconnecting of conductive parts to ensure
> common electrical potential between the bonded parts. Bonding is done pri-
> marily for two reasons. Conductive parts, when bonded using low impedance
> connections, would tend to be at the same electrical potential, meaning that
> the voltage difference between the bonded parts would be minimal or negli-
> gible. Bonding also ensures that any fault current likely imposed on a metal
> part will be safely conducted to ground or other grid systems serving as ground.

Capacitance — Property of a circuit element characterized by an insulating medium contained between two conductive parts. The unit of capacitance is a farad (F), named for the English scientist Michael Faraday. Capacitance values are more commonly expressed in microfarad (μF), which is 10^{-6} of a farad. Capacitance is one means by which energy or electrical noise can couple from one electrical circuit to another. Capacitance between two conductive parts can be made infinitesimally small but may not be completely eliminated.

Coupling — Process by which energy or electrical noise in one circuit can be transferred to another circuit that may or may not be electrically connected to it.

Crest factor — Ratio between the peak value and the root mean square (RMS) value of a periodic waveform. Figure 1.1 indicates the crest factor of two periodic waveforms. Crest factor is one indication of the distortion of a periodic waveform from its ideal characteristics.

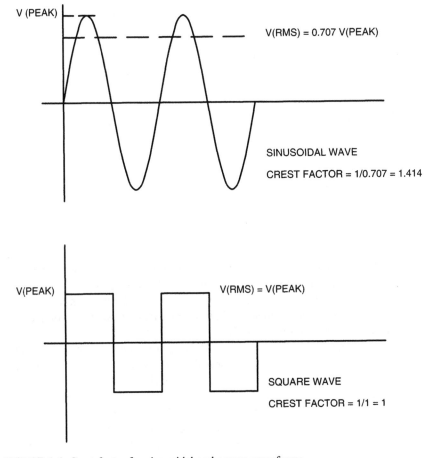

FIGURE 1.1 Crest factor for sinusoidal and square waveforms.

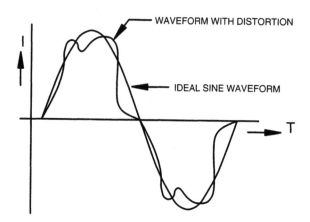

FIGURE 1.2 Waveform with distortion.

> **Distortion** — Qualitative term indicating the deviation of a periodic wave from its ideal waveform characteristics. Figure 1.2 contains an ideal sinusoidal wave along with a distorted wave. The distortion introduced in a wave can create waveform deformity as well as phase shift.
>
> **Distortion factor** — Ratio of the RMS of the harmonic content of a periodic wave to the RMS of the fundamental content of the wave, expressed as a percent. This is also known as the total harmonic distortion (THD); further explanation can be found in Chapter 4.
>
> **Flicker** — Variation of input voltage sufficient in duration to allow visual observation of a change in electric light source intensity. Quantitatively, flicker may be expressed as the change in voltage over nominal expressed as a percent. For example, if the voltage at a 120-V circuit increases to 125 V and then drops to 117 V, the flicker, f, is calculated as $f = 100 \times (125 - 117)/120 = 6.66\%$.
>
> **Form factor** — Ratio between the RMS value and the average value of a periodic waveform. Form factor is another indicator of the deviation of a periodic waveform from the ideal characteristics. For example, the average value of a pure sinusoidal wave averaged over a cycle is 0.637 times the peak value. The RMS value of the sinusoidal wave is 0.707 times the peak value. The form factor, FF, is calculated as $FF = 0.707/0.637 = 1.11$.
>
> **Frequency** — Number of complete cycles of a periodic wave in a unit time, usually 1 sec. The frequency of electrical quantities such as voltage and current is expressed in hertz (Hz).
>
> **Ground electrode** — Conductor or a body of conductors in intimate contact with earth for the purpose of providing a connection with the ground. Further explanation can be found in Chapter 5.
>
> **Ground grid** — System of interconnected bare conductors arranged in a pattern over a specified area and buried below the surface of the earth.
>
> **Ground loop** — Potentially detrimental loop formed when two or more points in an electrical system that are nominally at ground potential are connected by a conducting path such that either or both points are not at the same ground potential.

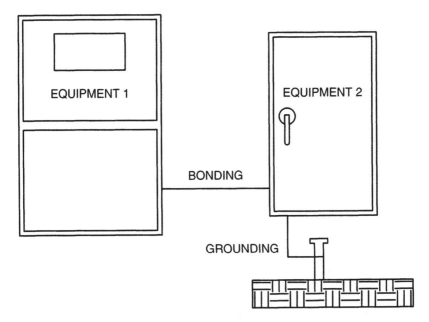

FIGURE 1.3 Bonding and grounding of equipment.

Ground ring — Ring encircling the building or structure in direct contact with the earth. This ring should be at a depth below the surface of the earth of not less than 2.5 ft and should consist of at least 20 ft of bare copper conductor not smaller than #2 AWG.

Grounding — Conducting connection by which an electrical circuit or equipment is connected to the earth or to some conducting body of relatively large extent that serves in place of the earth. In Figure 1.3, two conductive bodies are bonded and connected to ground. Grounding of metallic non-current-carrying parts of equipment is done primarily for safety reasons. Grounding the metal frame of equipment protects any person coming into contact with the equipment frame from electrical shock in case of a fault between an energized conductor and the frame. Grounding the equipment frame also ensures prompt passage of fault current to the ground electrode or ground plane; a protective device would operate to clear the fault and isolate the faulty equipment from the electrical power source.

Harmonic — Sinusoidal component of a periodic wave having a frequency that is an integral multiple of the fundamental frequency. If the fundamental frequency is 60 Hz, then the second harmonic is a sinusoidal wave of 120 Hz, the fifth harmonic is a sinusoidal wave of 300 Hz, and so on; see Chapter 4 for further discussion.

Harmonic distortion — Quantitative representation of the distortion from a pure sinusoidal waveform.

Impulse — Traditionally used to indicate a short duration overvoltage event with certain rise and fall characteristics. Standards have moved toward including the term *impulse* in the category of transients.

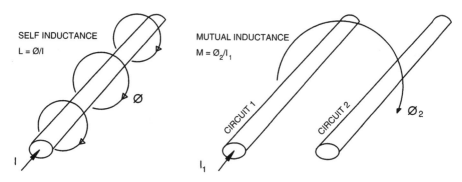

FIGURE 1.4 Self and mutual inductance.

Inductance — Inductance is the relationship between the magnetic lines of flux ($\emptyset$) linking a circuit due to the current (I) producing the flux. If I is the current in a wire that produces a magnetic flux of $\emptyset$ lines, then the self inductance of the wire, L, is equal to $\emptyset/I$. Mutual inductance (M) is the relationship between the magnetic flux $\emptyset_2$ linking an adjacent circuit 2 due to current I_1 in circuit 1. This can be stated as $M = \emptyset_2/I_1$. Figure 1.4 points out the two inductances. The unit of inductance is the henry [H], named for the American scientist Joseph Henry. The practical unit of inductance is the millihenry [mH], which is equal to 10^{-3} H. Self inductance of a circuit is important for determining the characteristics of impulse voltage transients and waveform notches. In power quality studies, we also are concerned with the mutual inductance as it relates to how current in one circuit can induce noise and disturbance in an adjacent circuit.

Inrush — Large current that a load draws when initially turned on.

Interruption — Complete loss of voltage or current for a time period.

Isolation — Means by which energized electrical circuits are uncoupled from each other. Two-winding transformers with primary and secondary windings are one example of isolation between circuits. In actuality, some coupling still exists in a two-winding transformer due to capacitance between the primary and the secondary windings.

Linear loads — Electrical load which in steady-state operation presents essentially constant impedance to the power source throughout the cycle of applied voltage. A purely linear load has only the fundamental component of the current present.

Noise — Electrical noise is unwanted electrical signals that produce undesirable effects in the circuits of control systems in which they occur. Figure 1.5 shows an example of noise in a 480-V power wiring due to switching resonance.

Nonlinear load — Electrical load that draws currents discontinuously or whose impedance varies during each cycle of the input AC voltage waveform. Figure 1.6 shows the waveform of a nonlinear current drawn by fluorescent lighting loads.

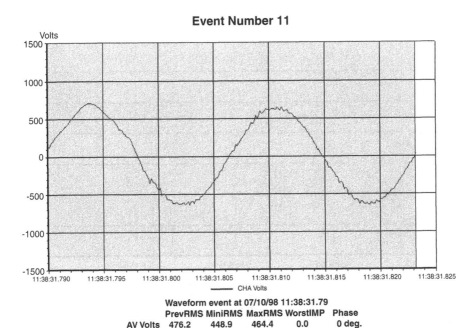

FIGURE 1.5 Noise in 480-V circuit due to switching resonance.

Notch — Disturbance of the normal power voltage waveform lasting less than a half cycle; the disturbance is initially of opposite polarity than the waveform and, thus, subtracts from the waveform. Figure 1.7 shows notch and noise produced by the operation of a converter in a variable speed drive.

Periodic — A voltage or current is periodic if the value of the function at time t is equal to the value at time $t + T$, where T is the period of the function. In this book, function refers to a periodic time-varying quantity such as AC voltage or current. Figure 1.8 is a periodic current waveform.

Power disturbance — Any deviation from the nominal value of the input AC characteristics.

Power factor (displacement) — Ratio between the active power (watts) of the fundamental wave to the apparent power (voltamperes) of the fundamental wave. For a pure sinusoidal waveform, only the fundamental component exists. The power factor, therefore, is the cosine of the displacement angle between the voltage and the current waveforms; see Figure 1.9.

Power factor (total) — Ratio of the total active power (watts) to the total apparent power (voltamperes) of the composite wave, including all harmonic frequency components. Due to harmonic frequency components, the total power factor is less than the displacement power factor, as the presence of harmonics tends to increase the displacement between the composite voltage and current waveforms.

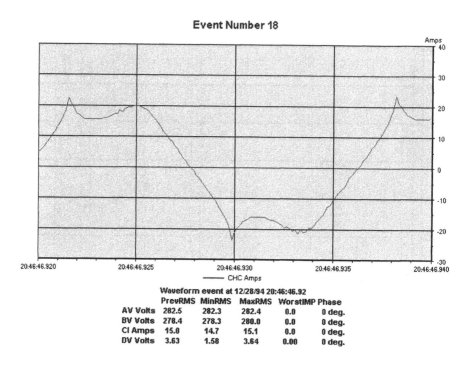

FIGURE 1.6 Nonlinear current drawn by fluorescent lighting loads.

Recovery time — Interval required for output voltage or current to return to a value within specifications after step load or line changes.

Ride through — Measure of the ability of control devices to sustain operation when subjected to partial or total loss of power of a specified duration.

Sag — RMS reduction in the AC voltage at power frequency from half of a cycle to a few seconds' duration. Figure 1.10 shows a sag lasting for 4 cycles.

Surge — Electrical transient characterized by a sharp increase in voltage or current.

Swell — RMS increase in AC voltage at power frequency from half of a cycle to a few seconds' duration. Figure 1.11 shows a swell of 2.5 cycles.

Transient — Subcycle disturbance in the AC waveform evidenced by a sharp, brief discontinuity of the waveform. This may be of either polarity and may be additive or subtractive from the nominal waveform. Transients occur when there is a sudden change in the voltage or the current in a power system. Transients are short-duration events, the characteristics of which are predominantly determined by the resistance, inductance, and capacitance of the power system network at the point of interest. The primary characteristics that define a transient are the peak amplitude, the rise time, the fall time, and the frequency of oscillation. Figure 1.12 shows a transient voltage waveform at the output of a power transformer as the result of switching-in of a motor containing power factor correction capacitors.

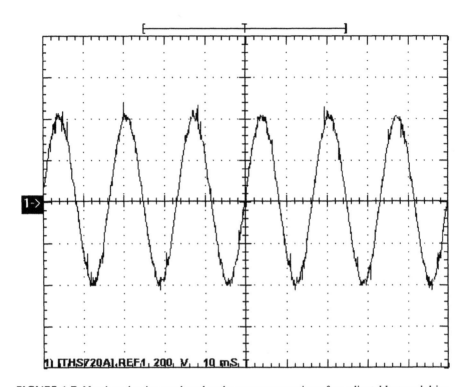

FIGURE 1.7 Notch and noise produced at the converter section of an adjustable speed drive.

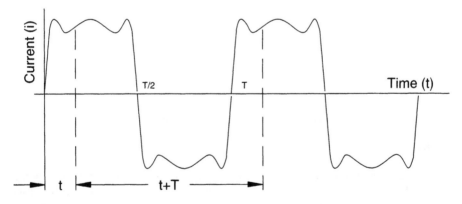

FIGURE 1.8 Periodic function of period *T.*

1.4 POWER QUALITY ISSUES

Power quality is a simple term, yet it describes a multitude of issues that are found in any electrical power system and is a subjective term. The concept of good and bad power depends on the end user. If a piece of equipment functions satisfactorily,

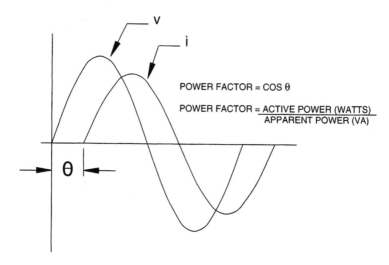

FIGURE 1.9 Displacement power factor.

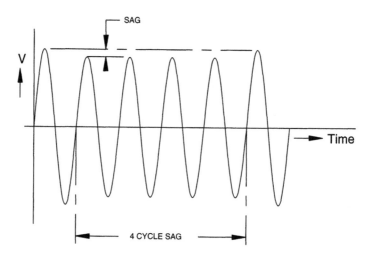

FIGURE 1.10 Voltage sag.

the user feels that the power is good. If the equipment does not function as intended or fails prematurely, there is a feeling that the power is bad. In between these limits, several grades or layers of power quality may exist, depending on the perspective of the power user. Understanding power quality issues is a good starting point for solving any power quality problem. Figure 1.13 provides an overview of the power quality issues that will be discussed in this book.

Power frequency disturbances are low-frequency phenomena that result in voltage sags or swells. These may be source or load generated due to faults or switching operations in a power system. The end results are the same as far as the susceptibility of electrical equipment is concerned. *Power system transients* are fast, short-duration

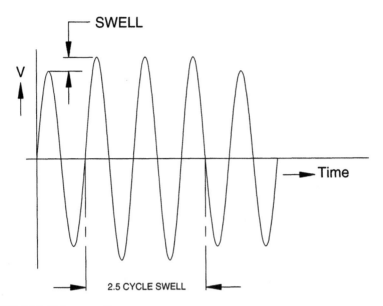

FIGURE 1.11 Voltage swell.

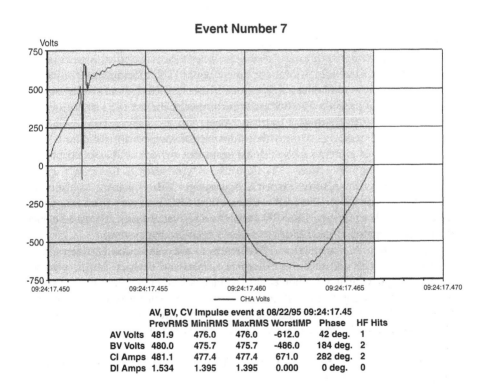

FIGURE 1.12 Motor starting transient voltage waveform.

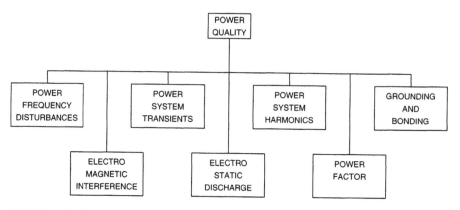

FIGURE 1.13 Power quality concerns.

events that produce distortions such as notching, ringing, and impulse. The mechanisms by which transient energy is propagated in power lines, transferred to other electrical circuits, and eventually dissipated are different from the factors that affect power frequency disturbances. *Power system harmonics* are low-frequency phenomena characterized by waveform distortion, which introduces harmonic frequency components. Voltage and current harmonics have undesirable effects on power system operation and power system components. In some instances, interaction between the harmonics and the power system parameters (R–L–C) can cause harmonics to multiply with severe consequences.

The subject of *grounding and bonding* is one of the more critical issues in power quality studies. Grounding is done for three reasons. The fundamental objective of grounding is safety, and nothing that is done in an electrical system should compromise the safety of people who work in the environment; in the U.S., safety grounding is mandated by the National Electrical Code (NEC®). The second objective of grounding and bonding is to provide a low-impedance path for the flow of fault current in case of a ground fault so that the protective device could isolate the faulted circuit from the power source. The third use of grounding is to create a ground reference plane for sensitive electrical equipment. This is known as the signal reference ground (SRG). The configuration of the SRG may vary from user to user and from facility to facility. The SRG cannot be an isolated entity. It must be bonded to the safety ground of the facility to create a total ground system.

Electromagnetic interference (EMI) refers to the interaction between electric and magnetic fields and sensitive electronic circuits and devices. EMI is predominantly a high-frequency phenomenon. The mechanism of coupling EMI to sensitive devices is different from that for power frequency disturbances and electrical transients. The mitigation of the effects of EMI requires special techniques, as will be seen later. *Radio frequency interference* (RFI) is the interaction between conducted or radiated radio frequency fields and sensitive data and communication equipment. It is convenient to include RFI in the category of EMI, but the two phenomena are distinct.

Electrostatic discharge (ESD) is a very familiar and unpleasant occurrence. In our day-to-day lives, ESD is an uncomfortable nuisance we are subjected to when we open the door of a car or the refrigerated case in the supermarket. But, at high levels, ESD is harmful to electronic equipment, causing malfunction and damage. *Power factor* is included for the sake of completing the power quality discussion. In some cases, low power factor is responsible for equipment damage due to component overload. For the most part, power factor is an economic issue in the operation of a power system. As utilities are increasingly faced with power demands that exceed generation capability, the penalty for low power factor is expected to increase. An understanding of the power factor and how to remedy low power factor conditions is not any less important than understanding other factors that determine the health of a power system.

1.5 SUSCEPTIBILITY CRITERIA

1.5.1 CAUSE AND EFFECT

The subject of power quality is one of cause and effect. Power quality is the cause, and the ability of the electrical equipment to function in the power quality environment is the effect. The ability of the equipment to perform in the installed environment is an indicator of its immunity. Figures 1.14 and 1.15 show power quality and equipment immunity in two forms. If the equipment immunity contour is within the power quality boundary, as shown in Figure 1.14, then problems can be expected. If the equipment immunity contour is outside the power quality boundary, then the equipment should function satisfactorily. The objective of any power quality study or solution is to ensure that the immunity contour is outside the boundaries of the power quality contour. Two methods for solving a power quality problem are to either make the power quality contour smaller so that it falls within the immunity contour or make the immunity contour larger than the power quality contour.

In many cases, the power quality and immunity contours are not two-dimensional and may be more accurately represented three-dimensionally. While the ultimate goal is to fit the power quality mass inside the immunity mass, the process is complicated because, in some instances, the various power quality factors making up the mass are interdependent. Changing the limits of one power quality factor can result in another factor falling outside the boundaries of the immunity mass. This concept is fundamental to solving power quality problems. In many cases, solving a problem involves applying multiple solutions, each of which by itself may not be the cure. Figure 1.16 is a two-dimensional immunity graph that applies to an electric motor. Figure 1.17 is a three-dimensional graph that applies to an adjustable speed drive module. As the sensitivity of the equipment increases, so does the complexity of the immunity contour.

1.5.2 TREATMENT CRITERIA

Solving power quality problems requires knowledge of which pieces or subcomponents of the equipment are susceptible. If a machine reacts adversely to a

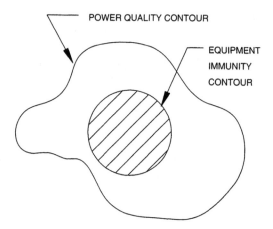

FIGURE 1.14 Criteria for equipment susceptibility.

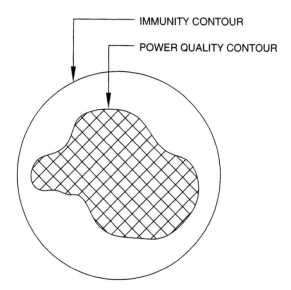

FIGURE 1.15 Criteria for equipment immunity.

particular power quality problem, do we try to treat the entire machine or treat the
subcomponent that is susceptible? Sometimes it may be more practical to treat the
subcomponent than the power quality for the complete machine, but, in other
instances, this may not be the best approach. Figure 1.18 is an example of treatment
of power quality at a component level. In this example, component A is susceptible
to voltage notch exceeding 30 V. It makes more sense to treat the power to
component A than to try to eliminate the notch in the voltage. In the same example,
if the power quality problem was due to ground loop potential, then component
treatment may not produce the required results. The treatment should then involve
the whole system.

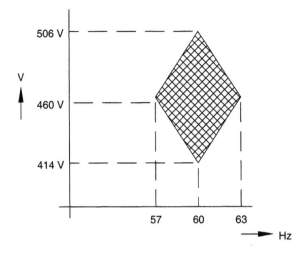

FIGURE 1.16 Volts–hertz immunity contour for 460-VAC motor.

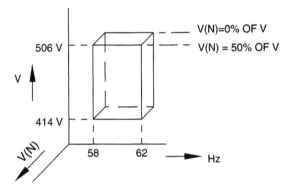

FIGURE 1.17 Volts–hertz–notch depth immunity contour for 460-V adjustable speed drive.

1.5.3 POWER QUALITY WEAK LINK

The reliability of a machine depends on the susceptibility of the component that has the smallest immunity mass. Even though the rest of the machine may be capable of enduring severe power quality problems, a single component can render the entire machine extremely susceptible. The following example should help to illustrate this.

A large adjustable speed drive in a paper mill was shutting down inexplicably and in random fashion. Each shutdown resulted in production loss, along with considerable time and expense to clean up the debris left by the interruption of production. Finally, after several hours of troubleshooting, the problem was traced to an electromechanical relay added to the drive unit during commissioning for a remote control function. This relay was an inexpensive, commercial-grade unit costing about $10. Once this relay was replaced, the drive operated satisfactorily

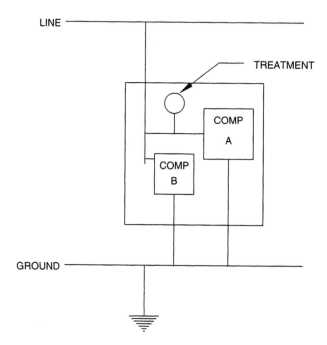

LINE

TREATMENT

COMP
A

COMP
B

GROUND

FIGURE 1.18 Localized power quality treatment.

without further interruptions. It is possible that a better grade relay would have prevented the shutdowns. Total cost of loss of production alone was estimated at $300,000. One does not need to look very far to see how important the weak link concept is when looking for power quality solutions.

1.5.4 INTERDEPENDENCE

Power quality interdependence means that two or more machines that could operate satisfactorily by themselves do not function properly when operating together in a power system. Several causes contribute to this occurrence. Some of the common causes are voltage fluctuations, waveform notching, ground loops, conducted or radiated electromagnetic interference, and transient impulses. In such a situation, each piece of equipment in question was likely tested at the factory for proper performance, but, when the pieces are installed together, power quality aberrations are produced that can render the total system inoperative. In some cases, the relative positions of the machines in the electrical system can make a difference. General guidelines for minimizing power quality interdependence include separating equipment that produces power quality problems from equipment that is susceptible. The offending machines should be located as close to the power source as possible. The power source may be viewed as a large pool of water. A disturbance in a large pool (like dropping a rock) sets out ripples, but these are small and quickly absorbed. As we move downstream from the power source, each location may be viewed as a smaller pool where any disturbance produces larger and longer-lasting ripples. At

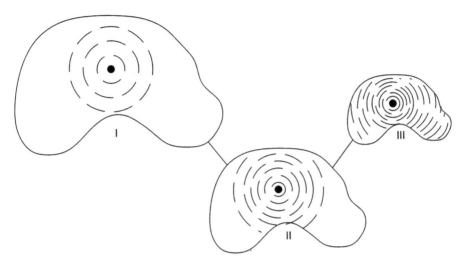

FIGURE 1.19 Power quality source dependence.

points farthest downstream from the source, even a small disturbance will have significant effects. Figure 1.19 illustrates this principle.

1.5.5 STRESS–STRAIN CRITERIA

In structural engineering, two frequently used terms are stress and strain. If load is applied to a beam, up to a point the resulting strain is proportional to the applied stress. The strain is within the elastic limit of the material of the beam. Loading beyond a certain point produces permanent deformity and weakens the member where the structural integrity is compromised. Electrical power systems are like structural beams. Loads that produce power quality anomalies can be added to a power system, to a point. The amount of such loads that may be tolerated depends on the rigidity of the power system. Rigid power systems can usually withstand a higher number of power quality offenders than weak systems. A point is finally reached, however, when further addition of such loads might make the power system unsound and unacceptable for sensitive loads. Figure 1.20 illustrates the stress–strain criteria in an electrical power system.

1.5.6 POWER QUALITY VS. EQUIPMENT IMMUNITY

All devices are susceptible to power quality; no devices are 100% immune. All electrical power system installations have power quality anomalies to some degree, and no power systems exist for which power quality problems are nonexistent. The challenge, therefore, is to create a balance. In Figure 1.21, the balanced beam represents the electrical power system. Power quality and equipment immunity are two forces working in opposition. The object is then to a create a balance between the two. We can assign power quality indices to the various locations in the power system and immunity indices to the loads. By matching the immunity index of a

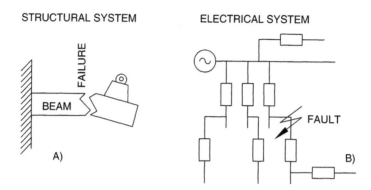

FIGURE 1.20 Structural and electrical system susceptibility.

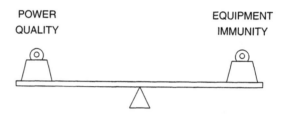

FIGURE 1.21 Power quality and equipment immunity.

piece of equipment with the power quality index, we can arrive at a balance where all equipment in the power system can coexist and function adequately. Experience indicates that three categories would sufficiently represent power quality and equipment immunity (see Table 1.1). During the design stages of a facility, many problems can be avoided if sufficient care is exercised to balance the immunity characteristics of equipment with the power quality environment.

1.6 RESPONSIBILITIES OF THE SUPPLIERS AND USERS OF ELECTRICAL POWER

The realization of quality electrical power is the responsibility of the suppliers and users of electricity. Suppliers are in the business of selling electricity to widely varying clientele. The needs of one user are usually not the same as the needs of other users. Most electrical equipment is designed to operate within a voltage of ±5% of nominal with marginal decrease in performance. For the most part, utilities are committed to adhering to these limits. At locations remote from substations supplying power from small generating stations, voltages outside of the ±5% limit are occasionally seen. Such a variance could have a negative impact on loads such as motors and fluorescent lighting. The overall effects of voltage excursions outside the nominal are not that significant unless the voltage approaches the limits of ±10% of nominal. Also, in urban areas, the utility frequencies are rarely outside ±0.1 Hz of the nominal frequency. This is well within the operating tolerance of most sensitive

TABLE 1.1
Immunity and Power Quality Indices

Index	Description	Examples
	Equipment Immunity Indices	
I	High immunity	Motors, transformers, incandescent lighting, heating loads, electromechanical relays
II	Moderate immunity	Electronic ballasts, solid-state relays, programmable logic controllers, adjustable speed drives
III	Low immunity	Signal, communication, and data processing equipment; electronic medical equipment
	Power Quality Indices	
I	Low power quality problems	Service entrance switchboard, lighting power distribution panel
II	Moderate power quality problems	HVAC power panels
III	High power quality problems	Panels supplying adjustable speed drives, elevators, large motors

equipment. Utilities often perform switching operations in electrical substations to support the loads. These can generate transient disturbances at levels that will have an impact on electrical equipment. While such transients generally go unnoticed, equipment failures due to these practices have been documented. Such events should be dealt with on a case-by-case basis. Figure 1.22 shows a 2-week voltage history for a commercial building. The nominal voltage at the electrical panel was 277 V phase to neutral. Two incidents of voltage sag can be observed in the voltage summary and were attributed to utility faults due to weather conditions. Figure 1.23 provides the frequency information for the same time period.

What are the responsibilities of the power consumer? Some issues that are relevant are energy conservation, harmonic current injection, power factor, and surge current demands. Given the condition that the utilities are becoming less able to keep up with the demand for electrical energy, it is incumbent on the power user to optimize use. Energy conservation is one means of ensuring an adequate supply of electrical power and at the same time realize an ecological balance. We are in an electronic age in which most equipment utilizing electricity generates harmonic-rich currents. The harmonics are injected into the power source, placing extra demands on the power generation and distribution equipment. As this trend continues to increase, more and more utilities are placing restrictions on the amount of harmonic current that the user may transmit into the power source.

The power user should also be concerned about power factor, which is the ratio of the real power (watts) consumed to the total apparent power (voltamperes) drawn from the source. In an ideal world, all apparent power drawn will be converted to useful work and supply any losses associated with performing the work. For several reasons, which will be discussed in a later chapter, this is not so in the real world. As the ratio between the real power needs of the system and the apparent power

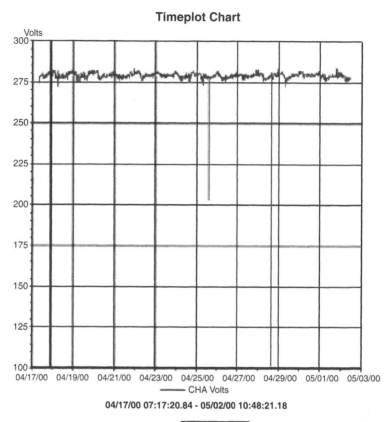

FIGURE 1.22 Voltage history graph at an electrical panel.

drawn from the source grows smaller, the efficiency with which power is being utilized is lowered. Typically, power suppliers expect a power factor of 0.95 or higher from industrial and commercial users of power. A penalty is levied if the power factor is below 0.95. Utilizing any one of several means, users can improve the power factor so the penalty may be avoided or minimized. It is not difficult to appreciate that if power suppliers and users each do their part, power quality is improved and power consumption is optimized.

1.7 POWER QUALITY STANDARDS

With the onset of the computer age and the increasing trend toward miniaturization of electrical and electronic devices, power quality problems have taken on increasing importance. The designers of computers and microprocess controllers are not versed in power system power quality issues. By the same token, power system designers and operators have limited knowledge of the operation of sensitive electronics. This environment has led to a need for power quality standards and guidelines. Currently,

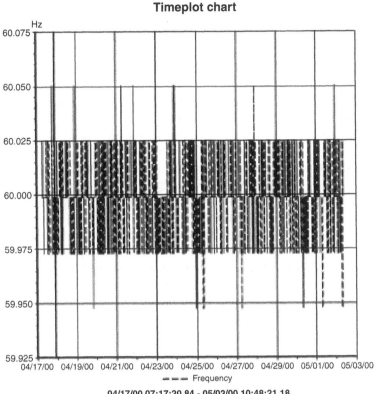

FIGURE 1.23 Frequency history at an electrical panel.

several engineering organizations and standard bearers in several parts of the world are spending a large amount of resources to generate power quality standards. Following is a list of power quality and related standards from two such organizations; some of the standards listed are in existence at this time, while others are still in process:

Institute of Electrical and Electronic Engineers (IEEE); Piscataway, NJ; http://www.ieee.org

IEEE 644	Standard Procedure for Measurement of Power Frequency Electric and Magnetic Fields from AC Power Lines
IEEE C63.12	Recommended Practice for Electromagnetic Compatibility Limits
IEEE 518	Guide for the Installation of Electrical Equipment to Minimize Electrical Noise Inputs to Controllers from External Sources

IEEE 519	Recommended Practices and Requirements for Harmonic Control in Electrical Power Systems
IEEE 1100	Recommended Practice for Powering and Grounding Sensitive Electronic Equipment
IEEE 1159	Recommended Practice for Monitoring Electric Power Quality
IEEE 141	Recommended Practice for Electric Power Distribution for Industrial Plants
IEEE 142	Recommended Practice for Grounding of Industrial and Commercial Power Systems
IEEE 241	Recommended Practice for Electric Power Systems in Commercial Buildings
IEEE 602	Recommended Practice for Electric Systems in Health Care Facilities
IEEE 902	Guide for Maintenance, Operation and Safety of Industrial and Commercial Power Systems
IEEE C57.110	Recommended Practice for Establishing Transformer Capability when Supplying Nonsinusoidal Load
IEEE P1433	Power Quality Definitions
IEEE P1453	Voltage Flicker
IEEE P1564	Voltage Sag Indices

International Electrotechnical Commission (IEC); Geneva, Switzerland; http://www.iec.ch

IEC/TR3 61000-2-1	Electromagnetic Compatibility — Environment
IEC/TR3 61000-3-6	Electromagnetic Compatibility — Limits
IEC 61000-4-7	Electromagnetic Compatibility — Testing and Measurement Techniques — General Guides on Harmonics and Interharmonics Measurements and Instrumentation
IEC 61642	Industrial a.c. Networks Affected by Harmonics — Application of Filters and Shunt Capacitors
IEC SC77A	Low Frequency EMC Phenomena
IEC TC77/WG1	Terminology
IEC SC77A/WG1	Harmonics and Other Low Frequency Disturbances
IEC SC77A/WG6	Low Frequency Immunity Tests
IEC SC77A/WG2	Voltage Fluctuations and Other Low Frequency Disturbances
IEC SC77A/WG8	Electromagnetic Interference Related to the Network Frequency
IEC SC77A/WG9	Power Quality Measurement Methods

1.8 CONCLUSIONS

The concept of power quality is a qualitative one for which mathematical expressions are not absolutely necessary to develop a basic understanding of the issues; however, mathematical expressions *are* necessary to solve power quality problems. If we cannot effectively represent a power quality problem with expressions based in mathematics, then solutions to the problem become exercises in trial and error. The expressions are what define power quality boundaries, as discussed earlier. So far, we have stayed away from much of quantitative analysis of power quality for the purpose of first developing an understanding. In later chapters, formulas and expressions will be introduced to complete the picture.

2 Power Frequency Disturbance

2.1 INTRODUCTION

The term *power frequency disturbance* describes events that are slower and longer lasting compared to electrical transients (see Chapter 3). Power frequency disturbances can last anywhere from one complete cycle to several seconds or even minutes. While the disturbance can be nothing more than an inconvenience manifesting itself as a flickering of lights or bumpy ride in an elevator, in other instances the effects can be harmful to electrical equipment. Typically, the deleterious effects of power frequency disturbances are predominantly felt in the long run, and such disturbances do not result in immediate failure of electrical devices.

The effects of power frequency disturbances vary from one piece of equipment to another and with the age of the equipment. Equipment that is old and has been subjected to harmful disturbances over a prolonged period is more susceptible to failure than new equipment. Fortunately, because power frequency disturbances are slower and longer lasting events, they are easily measured using instrumentation that is simple in construction.

2.2 COMMON POWER FREQUENCY DISTURBANCES

2.2.1 VOLTAGE SAGS

One of the most common power frequency disturbances is voltage sag. By definition, voltage sag is an event that can last from half of a cycle to several seconds. Voltage sags typically are due to starting on large loads, such as an electric motor or an arc furnace. Induction motors draw starting currents ranging between 600 and 800% of their nominal full load currents. The current starts at the high value and tapers off to the normal running current in about 2 to 8 sec, based on the motor design and load inertia. Depending on the instant at which the voltage is applied to the motor, the current can be highly asymmetrical.

Figure 2.1 contains the waveform of the starting current of a 50-hp induction motor with a rated full-load current of 60 A at 460 VAC. During the first half of the cycle, the asymmetrical current attains a peak value of 860 A. When the circuit feeding the motor has high impedance, appreciable voltage sag can be produced. Figure 2.2 shows a 100-kVA transformer feeding the 50-hp motor just described. If the transformer has a leakage reactance of 5.0%, the voltage sag due to starting this motor is calculated as follows:

Full load current of the 100-kVA transformer at 480 V = 120 A.
Voltage drop due to the starting inrush = 5.0 × 860 ÷ (120 × √2) = 25.3%.

If the reactance of the power lines and the utility transformer feeding this transformer were included in the calculations, the voltage sag would be worse than the value

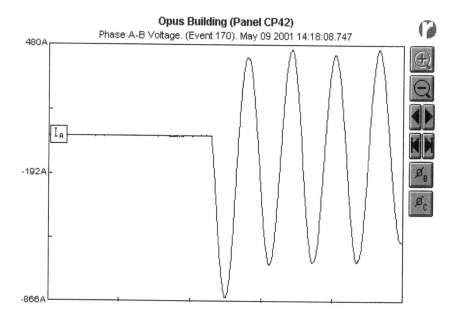

FIGURE 2.1 Motor-starting current waveform. A 5-hp motor was started across the line. The motor full-load current was 60 A. The first half-cycle peak reached a value of 860 A.

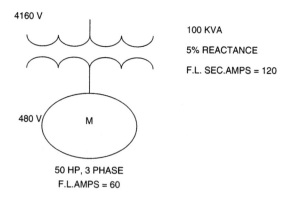

FIGURE 2.2 Schematic for example problem.

indicated. It is not difficult to see that any device that is sensitive to a voltage sag of 25% would be affected by the motor starting event.

Arc furnaces are another example of loads that can produce large voltage sags in electrical power systems. Arc furnaces operate by imposing a short circuit in a batch of metal and then drawing an arc, which produces temperatures in excess of 10,000°C, which melt the metal batch. Arc furnaces employ large inductors to stabilize the current due to the arc. Tens of thousands of amperes are drawn during the initial few seconds of the process. Figure 2.3 depicts typical current drawn by an arc furnace. Once the arc becomes stable, the current draw becomes more uniform. Due to the nature of the current drawn by the arc furnace, which is extremely nonlinear, large harmonic currents are also produced. Severe voltage sags are common in power lines that supply large arc furnaces, which are typically rated in the 30- to 50-MVA range and higher.

Arc furnaces are operated in conjunction with large capacitor banks and harmonic filters to improve the power factor and also to filter the harmonic frequency currents so they do not unduly affect other power users sharing the same power lines. It is not uncommon to see arc furnaces supplied from dedicated utility power lines to minimize their impact on other power users. The presence of large capacitance in an electrical system can result in voltage rise due to the leading reactive power demands of the capacitors, unless they are adequately canceled by the lagging reactive power required by the loads. This is why capacitor banks, whether for power factor correction or harmonic current filtration, are switched on when the furnace is brought on line and switched off when the arc furnace is off line.

Utility faults are also responsible for voltage sags. Approximately 70% of the utility-related faults occur in overhead power lines. Some common causes of utility faults are lightning strikes, contact with trees or birds and animals, and failure of insulators. The utility attempts to clear the fault by opening and closing the faulted circuit using reclosers, which can require from 40 to 60 cycles. The power line experiences voltage sags or total loss of power for the short duration it takes to clear the fault. Obviously, if the fault persists, the power outage continues until the problem

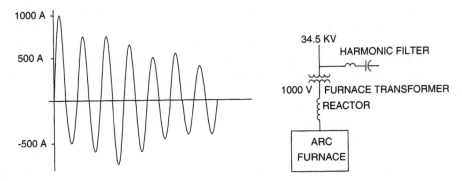

FIGURE 2.3 Typical current draw by arc furnace at the primary transformer. Large current fluctuations normally occur for several seconds before steady state is obtained.

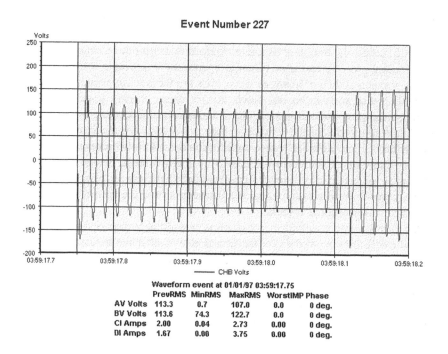

Event Number 227

Waveform event at 01/01/97 03:59:17.75

	PrevRMS	MinRMS	MaxRMS	WorstIMP	Phase
AV Volts	113.3	0.7	107.0	0.0	0 deg.
BV Volts	113.6	74.3	122.7	0.0	0 deg.
CI Amps	2.00	0.04	2.73	0.00	0 deg.
DI Amps	1.67	0.00	3.75	0.00	0 deg.

FIGURE 2.4 Voltage sag at a refinery due to a utility fault. The sag caused the programmable logic controller to drop out, which resulted in interruption of power. The sag lasted for approximately 21 cycles.

is corrected. Figure 2.4 shows voltage sag due to a utility fault near a refinery. This fault was attributed to stormy weather conditions just prior to the event. The sag lasted for approximately 21 cycles before the conditions returned to normal. The duration of the sag was sufficient to affect the operation of very critical process controllers in the refinery, which interrupted refinery production.

Figure 2.5 indicates voltage sag caused by utility switching operations near an aluminum smelter. Even though the sag lasted for only five cycles in this particular case, its magnitude was sufficient to cause several motor controllers to drop offline. Typically, electromagnetic controllers could ride through such events due to energy stored in their magnetic fields; however, the motor controllers in this facility contained electronic voltage-sensing circuits more sensitive to the voltage sags. In such cases, either the sensitive circuit should be adjusted to decrease its sensitivity or voltage-support devices such as capacitors or batteries should be provided.

Voltage sags and swells are also generated when loads are transferred from one power source to another. One example is transfer of load from the utility source to the standby generator source during loss of utility power. Most facilities contain emergency generators to maintain power to critical loads in case of an emergency. Sudden application and rejection of loads to a generator could create significant voltage sags or swells. Figures 2.6 and 2.7 show generator bus voltage during two sets of operating conditions. If critical loads are not able to withstand the imposed

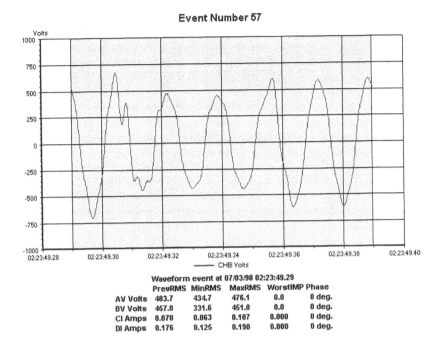

Event Number 57

Waveform event at 07/03/98 02:23:49.29

	PrevRMS	MinRMS	MaxRMS	WorstIMP	Phase
AV Volts	483.7	434.7	476.1	0.0	0 deg.
BV Volts	457.0	331.6	451.0	0.0	0 deg.
CI Amps	0.070	0.063	0.107	0.000	0 deg.
DI Amps	0.176	0.125	0.190	0.000	0 deg.

FIGURE 2.5 Voltage sag caused by utility switching at an aluminum smelter. The sag lasted for five cycles and caused motor controllers to drop out.

voltage conditions, problems are imminent. Such determinations should be made at the time when the generators are commissioned into service. The response of the generator or the sensitivity of the loads, or a combination of each, should be adjusted to obtain optimum performance. During power transfer from the utility to the generator, frequency deviations occur along with voltage changes. The generator frequency can fluctuate as much as ±5 Hz for a brief duration during this time. It is once again important to ensure that sensitive loads can perform satisfactorily within this frequency tolerance for the duration of the disturbance.

Flicker (or light flicker, as it is sometimes called) is a low-frequency phenomenon in which the magnitude of the voltage or frequency changes at such a rate as to be perceptible to the human eye. Flicker is also a subjective term. Depending on the sensitivity of the observer, light flicker may or may not be perceived. For instance, it has been found that 50% of people tested perceive light flicker as an annoyance under the following conditions:

Voltage Change	Changes/Second
1 volt	4
2 volts	2
4 volts	1

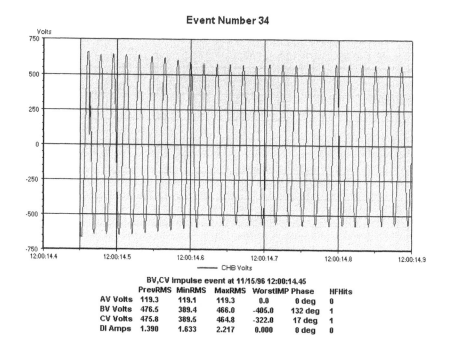

FIGURE 2.6 Voltage sag due to generator step load application. The nominal 480-V generator bus experienced a sag to 389 V that lasted for approximately 1 sec.

For small voltage changes that occur infrequently, flicker is not a serious concern but certainly can be a source of major annoyance. Typically, flicker is caused when a load that requires large currents during startup is initially energized. If the starts are frequent or if the current requirement of the load fluctuates rapidly during each cycle of operation, then flicker effects can be quite pronounced. Examples of loads that could cause light flicker are elevators, arc furnaces, and arc welders.

Figure 2.8 shows the voltage at a lighting panel supplying a multiunit residence. The graph shows the voltage change when one of the elevators for the building is operated, during which the voltage at the panel changes at the rate of 3 to 4 V. The light flicker in the building was quite evident. The problem in this particular building was due to one switchboard feeding both the lighting panel and the elevators. Under these conditions, unless the source supplying the load is large, interaction between the elevator and the lighting is likely to occur.

Flicker is expressed as:

$$f_v = 100 \times (V_{max} - V_{min})/V_{nom}$$

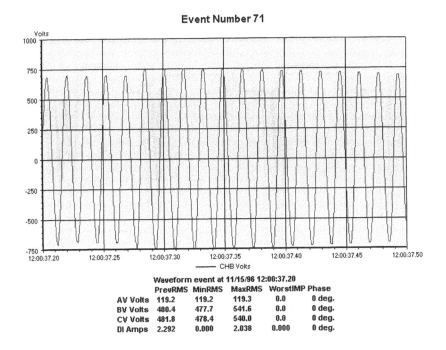

FIGURE 2.7 Voltage swell due to step load rejection. The nominal 480-V generator bus experienced a rise to 541 V that lasted for approximately 18 cycles.

where V_{max} and V_{min} represent the change in voltage over the nominal voltage V_{nom}. For example, if the voltage in a circuit rated at 120 V nominal changed from 122 to 115 V, the flicker is given by:

$$f_v = 100 \times (122 - 115)/120 = 5.83\%$$

In the early stages of development of AC power, light flicker was a serious problem. Power generation and distribution systems were not stiff enough to absorb large fluctuating currents. Manufacturing facilities used a large number of pumps and compressors of reciprocating design. Due to their pulsating power requirements, light flicker was a frequent problem. The use of centrifugal- or impeller-type pumps and compressors reduced the flicker problem considerably. The flicker problems were not, for the most part, eliminated until large generating stations came into service.

Light flicker due to arc furnaces requires extra mention. Arc furnaces, commonly found in many industrial towns, typically use scrap metal as the starting point. An arc is struck in the metal by applying voltage to the batch from a specially constructed furnace transformer. The heat due to the arc melts the scrap metal, which is drawn out from the furnace to produce raw material for a variety of industrial facilities. Arc furnaces impose large electrical power requirements on the electrical system.

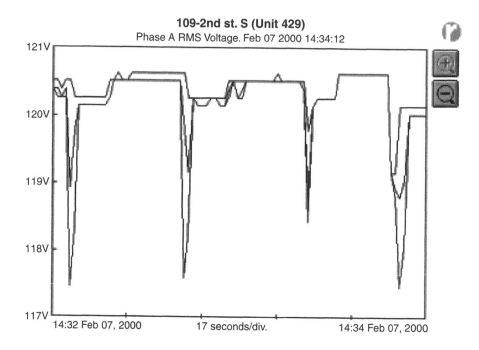

FIGURE 2.8 Voltage changes during elevator operation in a residential multiunit complex. The rate of voltage change causes perceptible light flicker.

The current drawn from the source tends to be highly cyclic as arcs are repeatedly struck and stabilized in different parts of the batch. The voltage at the supply lines to an arc furnace might appear as shown in Figure 2.9. The envelope of the change in voltage represents the flicker content of the voltage. The rate at which the voltage changes is the flicker frequency:

$$\Delta V = V_{max} - V_{min}$$

$$V_{nom} = \text{average voltage} = (V_{max} + V_{min})/2$$

$$f = 2 \times (V_{max} - V_{min}) \times 100/(V_{max} + V_{min})$$

Normally, we would use root mean square (RMS) values for the calculations, but, assuming that the voltages are sinusoidal, we could use the maximum values and still derive the same results. It has been found that a flicker frequency of 8 to 10 Hz with a voltage variation of 0.3 to 0.4% is usually the threshold of perception that leads to annoyance.

Arc furnaces are normally operated with capacitor banks or capacitor bank/filter circuits, which can amplify some of the characteristic frequency harmonic currents generated by the furnace, leading to severe light flicker. For arc furnace

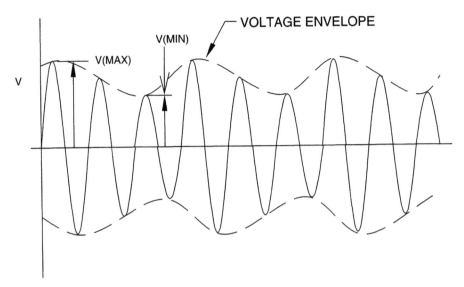

FIGURE 2.9 Typical arc furnace supply voltage indicating voltage fluctuation at the flicker frequency.

applications, careful planning is essential in the configuration and placement of the furnace and the filters to minimize flicker. Very often, arc furnaces are supplied by dedicated utility power lines that are not shared by other users. This follows from the principle that as the voltage source becomes larger (lower source impedance), the tendency to produce voltage flicker due to the operation of arc furnaces is lessened.

Low-frequency noise superimposed on the fundamental power frequency is a power quality concern. Discussion of this phenomenon is included in this chapter mainly because these are slower events that do not readily fit into any other category. Low-frequency noise is a signal with a frequency that is a multiple of the fundamental power frequency. Figure 2.10 illustrates a voltage waveform found in an aluminum smelting plant. In this plant, when the aluminum pot lines are operating, power factor improvement capacitors are also brought online to improve the power factor. When the capacitor banks are online, no significant noise is noticed in the power lines. When the capacitor banks are turned off, noise can be found on the voltage waveform (as shown) because the capacitor banks absorb the higher order harmonic frequency currents produced by the rectifiers feeding the pot lines. In this facility, the rest of the power system is not affected by the noise because of the low magnitudes. It is conceivable that at higher levels the noise could couple to nearby signal or communication circuits and cause problems.

Adjustable speed drives (ASDs) produce noise signals that are very often troublesome. The noise frequency generated by the ASDs is typically higher than the harmonic frequencies of the fundamental voltage. Because of this, the noise could find its way into sensitive data and signal circuits unless such circuits are sufficiently isolated from the ASD power lines.

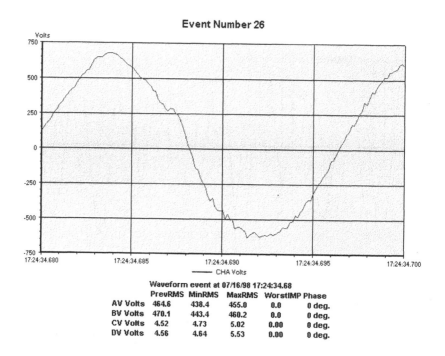

Event Number 26

	PrevRMS	MinRMS	MaxRMS	WorstIMP	Phase
AV Volts	464.6	438.4	455.0	0.0	0 deg.
BV Volts	470.1	443.4	460.2	0.0	0 deg.
CV Volts	4.52	4.73	5.02	0.00	0 deg.
DV Volts	4.56	4.64	5.53	0.00	0 deg.

Waveform event at 07/16/98 17:24:34.68

FIGURE 2.10 Low-frequency noise superimposed on the 480-V bus after switching off the capacitor bank.

2.3 CURES FOR LOW-FREQUENCY DISTURBANCES

Power-frequency or low-frequency disturbances are slow phenomena caused by switching events related to the power frequency. Such disturbances are dispersed with time once the incident causing the disturbance is removed. This allows the power system to return to normal operation. Low-frequency disturbances also reveal themselves more readily. For example, dimming of lights accompanies voltage sag on the system; when the voltage rises, lights shine brighter. While low-frequency disturbances are easily detected or measured, they are not easily corrected. Transients, on the other hand, are not easily detected or measured but are cured with much more ease than a low-frequency event. Measures available to deal with low-frequency disturbances are discussed in this section.

2.3.1 ISOLATION TRANSFORMERS

Isolation transformers, as their name indicates, have primary and secondary windings, which are separated by an insulating or isolating medium. Isolation transformers do not help in curing voltage sags or swells; they merely transform the voltage from a primary level to a secondary level to enable power transfer from one winding to the other. However, if the problem is due to common mode noise, isolation transformers help to minimize noise coupling, and shielded isolation transformers

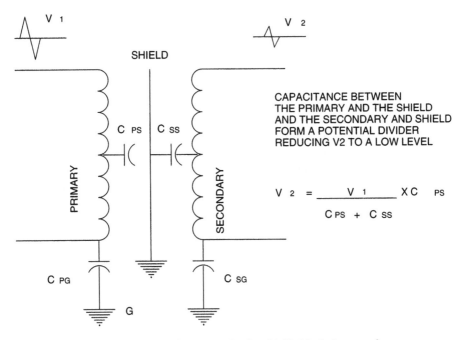

FIGURE 2.11 Common mode noise attenuation by shielded isolation transformer.

can help to a greater degree. Common mode noise is equally present in the line and the neutral circuits with respect to ground. Common mode noise may be converted to transverse mode noise (noise between the line and the neutral) in electrical circuits, which is troublesome for sensitive data and signal circuits. Shielded isolation transformers can limit the amount of common mode noise converted to transverse mode noise. The effectiveness with which a transformer limits common mode noise is called *attenuation* (A) and is expressed in decibels (dB):

$$A = 20 \log (V_1/V_2)$$

where V_1 is the common mode noise voltage at the transformer primary and V_2 is the differential mode noise at the transformer secondary. Figure 2.11 shows how common mode noise attenuation is obtained by the use of a shielded isolation transformer. The presence of a shield between the primary and secondary windings reduces the interwinding capacitance and thereby reduces noise coupling between the two windings.

Example: Find the attenuation of a transformer that can limit 1 V common mode noise to 10 mV of transverse mode noise at the secondary:

$$A = 20 \log (1/0.01) = 40 \text{ dB}$$

Isolation transformers using a single shield provide attenuation in a range of 40 to 60 dB. Higher attenuation may be obtained by specially designed isolation transformers using multiple shields configured to form a continuous enclosure around the secondary winding. Attenuation of the order of 100 dB may be realized with such techniques.

2.3.2 VOLTAGE REGULATORS

Voltage regulators are devices that can maintain a constant voltage (within tolerance) for voltage changes of predetermined limits above and below the nominal value. A switching voltage regulator maintains constant output voltage by switching the taps of an autotransformer in response to changes in the system voltage, as shown in Figure 2.12. The electronic switch responds to a signal from the voltage-sensing circuitry and switches to the tap connection necessary to maintain the output voltage constant. The switching is typically accomplished within half of a cycle, which is within the ride-through capability of most sensitive devices.

Ferro-resonant voltage regulators are static devices that have no moving components. They operate on the principle that, in a transformer, when the secondary magnetic circuit is operating in the saturation region the secondary winding is decoupled from the primary and therefore is not sensitive to voltage changes in the primary. The secondary winding has a capacitor connected across its terminals that

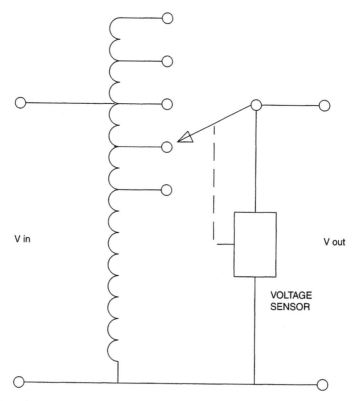

FIGURE 2.12 Tap-changer voltage regulator.

forms a parallel resonant circuit with the inductance of the secondary winding. Large magnetic fields are created in the magnetic core surrounding the secondary windings, thereby decoupling the secondary winding from the primary. Typically ferro-resonant transformer regulators can maintain secondary voltage to within ±0.5% for changes in the primary voltages of ±20%. Figure 2.13 contains the schematic of a ferro-resonance transformer type of voltage regulator.

Ferro-resonance transformers are sensitive to loads above their rated current. In extreme cases of overload, secondary windings can become detuned, at which point the output of the transformer becomes very low. Voltage sags far below the rated level can also have a detuning effect on the transformer. Within the rated voltage and load limits, however, the ferro-resonance transformer regulators are very effective in maintaining fairly constant voltage levels.

2.3.3 STATIC UNINTERRUPTIBLE POWER SOURCE SYSTEMS

Static uninterruptible power sources (UPSs) have no rotating parts, such as motors or generators. These are devices that maintain power to the loads during loss of

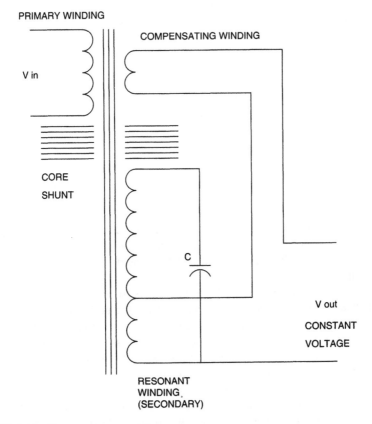

FIGURE 2.13 Ferro-resonant transformer.

normal power for a duration that is a function of the individual UPS system. All UPS units have an input rectifier to convert the AC voltage into DC voltage, a battery system to provide power to loads during loss of normal power, and an inverter which converts the DC voltage of the battery to an AC voltage suitable for the load being supplied. Depending on the UPS unit, these three main components are configured differently. Static UPS systems may be broadly classified into offline and online units. In the offline units, the loads are normally supplied from the primary electrical source directly. The primary electrical source may be utility power or an in-house generator. If the primary power source fails or falls outside preset parameters, the power to the loads is switched to the batteries and the inverter. The switching is accomplished within half of a cycle in most UPS units, thereby allowing critical loads to continue to receive power. During power transfer from the normal power to the batteries, the loads might be subjected to transients. Once the loads are transferred to the batteries, the length of time for which the loads would continue to receive power depends on the capacity of the batteries and the amount of load. UPS units usually can supply power for 15 to 30 min, at which time the batteries become depleted to a level insufficient to supply the loads, and the UPS unit shuts down. Some offline UPS system manufacturers provide optional battery packs to enhance the time of operation of the units after loss of normal power.

In online UPS units, normal power is rectified into DC power and in turn inverted to AC power to supply the loads. The loads are continuously supplied from the DC bus even during times when the normal power is available. A battery system is also connected to the DC bus of the UPS unit and kept charged from the normal source. When normal power fails, the DC bus is supplied from the battery system. No actual power transfer occurs during this time, as the batteries are already connected to the DC bus. Online units can be equipped with options such as manual and static bypass switches to circumvent the UPS and supply power to the loads directly from the normal source or an alternate source such as a standby generator. An offline unit is shown in Figure 2.14, and an online unit in Figure 2.15. Two important advantages of online UPS units are because: (1) power is normally supplied from the DC bus, the UPS unit in effect isolates the loads from the source which keeps power system disturbances and transients from interacting with the loads, and (2) since power to the loads is not switched during loss of normal power, no switching transients are produced. As might be expected, online UPS systems cost considerably more than offline units.

The output voltage of static UPS units tends to contain waveform distortions higher than those for normal power derived from the utility or a generator. This is due to the presence of the inverter in the output section of the UPS system. For some lower priced UPS units, the distortion can be substantial, with the waveform resembling a square wave. Figure 2.16 shows the output waveform of a UPS unit commonly used in offices to supply computer workstations. More expensive units use higher order inverter sections to improve the waveform of the output voltage, as shown in Figure 2.17. It is important to take into consideration the level of susceptibility of the loads to waveform distortion. Problems attributed to excessive voltage distortion have been noticed in some applications involving medical electronics and voice communication.

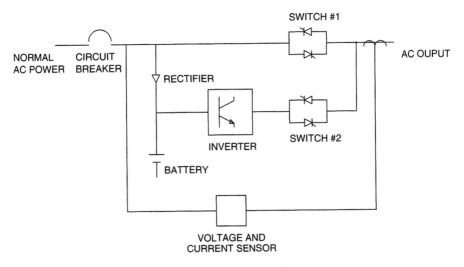

FIGURE 2.14 Offline uninterruptible power source (UPS) system.

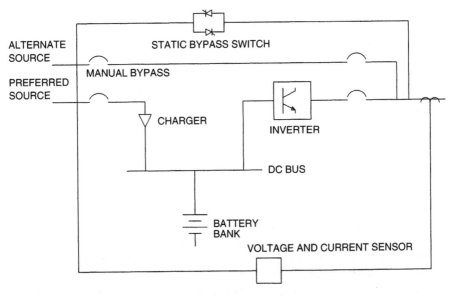

FIGURE 2.15 Online uninterruptible power source (UPS) system.

2.3.4 Rotary Uninterruptible Power Source Units

Rotary UPS (RUPS) units utilize rotating members to provide uninterrupted power
to loads, as shown in Figure 2.18. In this configuration, an AC induction motor
drives an AC generator, which supplies power to critical loads. The motor operates
from normal utility power. A diesel engine or other type of prime mover is coupled
to the same shaft as the motor and the generator. During normal operation, the diesel
engine is decoupled from the common shaft by an electric clutch. If the utility power

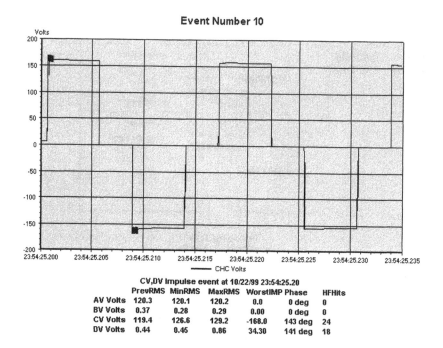

FIGURE 2.16 Output voltage waveform from an offline uninterruptible power source (UPS) system. The ringing during switching is evident during the first cycle.

fails, the prime mover shaft is coupled to the generator shaft and the generator gets its mechanical power from the prime mover. The motor shaft is attached to a flywheel, and the total inertia of the system is sufficient to maintain power to the loads until the prime mover comes up to full speed. Once the normal power returns, the induction motor becomes the primary source of mechanical power and the prime mover is decoupled from the shaft.

In a different type of RUPS system, during loss of normal power the AC motor is supplied from a battery bank by means of an inverter (Figure 2.19). The batteries are kept charged by the normal power source. The motor is powered from the batteries until the batteries become depleted. In some applications, standby generators are used to supply the battery bank in case of loss of normal power. Other combinations are used to provide uninterrupted power to critical loads, but we will not attempt to review all the available technologies. It is sufficient to point out that low-frequency disturbances are effectively mitigated using one of the means mentioned in this section.

2.4 VOLTAGE TOLERANCE CRITERIA

Manufacturers of computers and data-processing equipment do not generally publish data informing the user of the voltage tolerance limits for their equipment. An agency

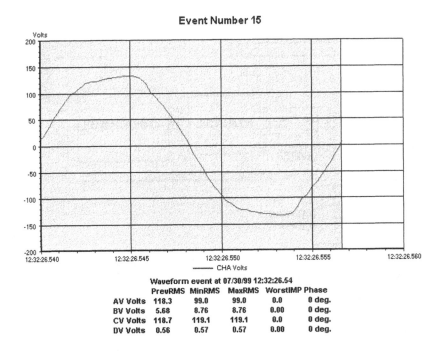

Event Number 15

Waveform event at 07/30/99 12:32:26.54

	PrevRMS	MinRMS	MaxRMS	WorstIMP	Phase
AV Volts	118.3	99.0	99.0	0.0	0 deg.
BV Volts	5.68	8.76	8.76	0.00	0 deg.
CV Volts	118.7	119.1	119.1	0.0	0 deg.
DV Volts	0.56	0.57	0.57	0.00	0 deg.

FIGURE 2.17 Voltage waveform from an online uninterruptible power source (UPS) system. The waveform, even though less than ideal, contains considerably lower distortion than the waveform of the offline unit shown in Figure 2.16.

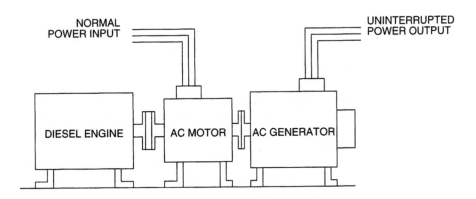

FIGURE 2.18 Rotary uninterruptible power source (RUPS) system using a diesel engine, AC motor, and AC generator to supply uninterrupted power to critical loads.

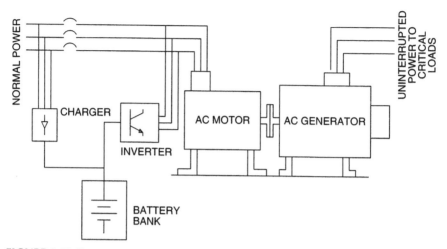

FIGURE 2.19 Rotary uninterruptible power source (RUPS) system using a battery bank, AC motor, and AC generator to provide uninterrupted power to critical loads.

known as the Information Technology Industry Council (ITIC) has published a graph that provides guidelines as to the voltage tolerance limits within which information technology equipment should function satisfactorily (Figure 2.20). The ordinate (y-axis) represents the voltage as a percentage of the nominal voltage. The abscissa (x-axis) is the time duration in seconds (or cycles). The graph contains three regions. The area within the graph is the voltage tolerance envelope, in which equipment should operate satisfactorily. The area above the graph is the prohibited region, in which equipment damage might result. The area below the graph is the region where the equipment might not function satisfactorily but no damage to the equipment should result. Several types of events fall within the regions bounded by the ITIC graph, as described below:

- *Steady-State Tolerance.* The steady-state range describes an RMS voltage that is either slowly varying or is constant. The subject range is ±10% from the nominal voltage. Any voltage in this range may be present for an indefinite period and is a function of the normal loading and losses in the distribution system.
- *Line Voltage Swell.* This region describes a voltage swell having an RMS amplitude up to 120% of the nominal voltage, with a duration of up to 0.5 sec. This transient may occur when large loads are removed from the system or when voltage is applied from sources other than the utility.
- *Low-Frequency Decaying Ring Wave.* This region describes a decaying ring wave transient that typically results from the connection of power factor correction capacitors to an AC power distribution system. The frequency of this transient may vary from 200 Hz to 5 kHz, depending on the resonant frequency of the AC distribution system. The magnitude of the transient is expressed as a percentage of the peak 60 Hz nominal (not the RMS). The transient is assumed to be completely decayed by the end of the half-cycle in which it occurs. The transient is assumed to occur

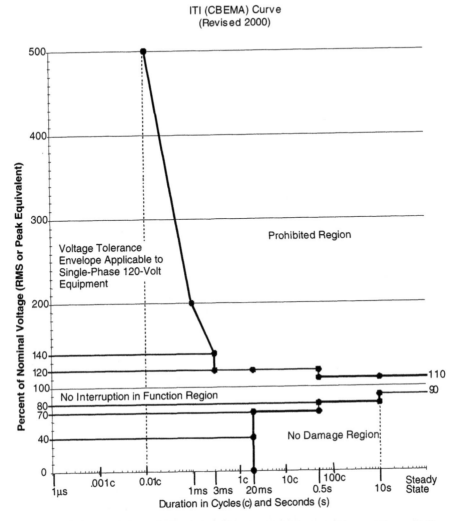

ITI (CBEMA) Curve
(Revised 2000)

FIGURE 2.20 Information Technology Industry Council (ITIC) graph providing guidelines as to the voltage tolerance limits within which information technology equipment should function satisfactorily. (Courtesy of the Information Technology Industry Council, Washington, D.C.)

near the peak of the nominal voltage waveform. The amplitude of the transient varies from 140% for 200-Hz ring waves to 200% for 5-kHz ring waves, with a linear increase in amplitude with frequency.

- *High-Frequency Impulse Ring Wave.* This region describes the transients that typically occur as the result of lightning strikes. Waveshapes applicable to this transient and general test conditions are described in the ANSI/IEEE C62.41 standard. This region of the curve deals with both amplitude and duration (energy) rather than RMS amplitude. The intent is to provide 80 J minimum transient immunity.

- *Voltage Sags.* Two different RMS voltage sags are described. Generally the transients result from application of heavy loads as well as fault conditions at various points in the AC power distribution system. Sags to 80% of nominal are assumed to have a typical duration of up to 10 sec and sags to 70% of nominal are assumed to have a duration of up to 0.5 sec.
- *Drop Out.* Voltage drop out includes both severe RMS voltage sags and complete interruption of the applied voltage followed by immediate reapplication of the nominal voltage. The interruption may last up to 20 msec. The transient typically results from the occurrence and subsequent clearing of the faults in the distribution system.
- *No Damage Region.* Events in this region include sags and drop outs that are more severe than those specified in the preceding paragraphs and continuously applied voltages that are less than the lower limit of the steady-state tolerance range. A normal functional state of the information technology equipment is not expected during these conditions, but no damage to equipment should result.
- *Prohibited Region.* This region includes any surge or swell which exceed the upper limit of the envelope. If information technology equipment is subjected to such conditions damage might result.

The ITIC graph apples to 120-V circuits obtained from 120-V, 120/240-V, and 120/208-V distribution systems. Other nominal voltages and frequencies are not specifically considered, but their applicability may be determined in each case. The curve is useful in determining if problems could be expected under particular power system voltage conditions (see Chapter 9).

2.5 CONCLUSIONS

Power frequency disturbances are perhaps not as damaging to electrical equipment as short time transients, but they can cause a variety of problems in the operation of an electrical power system. These disturbances may be utility (source) generated or generated within a facility due to the loads. Disturbances propagated from the source are not easily cured and fixed because, at the source level, we may be dealing with very high power and energy levels and the cures and fixes tend to be complex and expensive. However, disturbances internal to the facility are more easily cured or controlled. The effects of a disturbance within the facility may be minimized by separating the offending loads from the sensitive, susceptible loads. The offending loads should be located as close to the source of electrical power as possible to minimize their impact on the rest of the power system. Whether the power frequency disturbances are internal or external to a facility, the power conditioners discussed here are effective in dealing with these events.

3   Electrical Transients

3.1   INTRODUCTION

In Chapter 1, a *transient* is defined as a subcycle disturbance in the AC waveform that is discernible as a sharp discontinuity of the waveform. The definition states that transients are subcycle events, lasting less than one cycle of the AC waveform. Inclusion of the term *subcycle* is for the sake of definition only. Routinely we see transients that span several cycles. To satisfy the absolute definition, the transient occurring in the next cycle is not considered an extension of the transient in the previous cycle. This approach allows us to isolate the disturbance on a cycle-by-cycle basis for ease of analysis and treatment.

Subcycle transients are some of the most difficult anomalies to detect and treat. Their occurrence can be random, and they can vary in degree depending on the operating environment at the time of occurrence. Their effect on devices varies depending on the device itself and its location in an electrical system. Transients are difficult to detect because of their short duration. Conventional meters are not able to detect or measure them due to their limited frequency response or sampling rate. For example, if a transient occurs for 2 msec and is characterized by a frequency content of 20 kHz, the measuring instrument must have a frequency response or sampling rate of at least 10 times 20 kHz, or 200 kHz, in order to fairly describe the characteristics of the transient. For faster transients, higher sampling rates are necessary. In Chapter 9, we will look into what is involved in selecting and setting up instrumentation for measuring electrical transients.

Many different terms are associated with transients, such as *spikes*, *bumps*, *power pulses*, *impulses*, and *surges*. While some of these terms may indeed describe a particular transient, such terms are not recommended due to their ambiguity. We will use the term *transient* to denote all subcycle events and ascribe certain characteristics to these events, such as overvoltage, notch, noise, and ring. It may be that what we call transient is not as important as remembering that the underlying causes of the various transient events differ, as do the cures for the ill effects produced by the transients.

Why is an understanding of the transient phenomenon important? Large electromagnetic devices such as transformers and motors are practically impervious to the effects of transients. Problems arise because of the sensitivity of the microelectronic devices and circuits that make up the control elements of the power system. The microprocess controller is the nerve center of every present-day manufacturing or commercial facility. Medical electronic instruments used in healthcare facilities are becoming more sophisticated and at the same time increasingly susceptible to electrical transients. Because the performance of any machine is only as good as its weakest link, expansive operations can be rendered vulnerable due to the susceptibility of the most inexpensive and seemingly insignificant of the components comprising the system. We will examine some specific examples of such vulnerability later in this chapter.

3.2 TRANSIENT SYSTEM MODEL

Steady-state systems are the opposite of transient systems. In steady state, the operation of a power system is characterized by the fundamental frequency or by some low-frequency harmonic of the fundamental frequency. The three passive parameters of the system — resistance (R), inductance (L), and capacitance (C) — determine how the steady-state system will respond when under the influence of an applied voltage. This situation is somewhat analogous to the steady-state thermodynamics model, in which thermal energy supplied is equal to energy stored plus energy dissipated, and by using traditional laws of physics the temperature rise of devices can be accurately calculated. However, in a transient state, traditional laws of equilibrium do not apply. The circuit model of a transient electrical system will appear considerably different from the steady-state model. Passive parameters R, L, and C are still major determinants of the transient response, but their effect on the transient can change with the duration of the transient. In an electrical system, inductance and capacitance are the energy-storing elements that contribute to the oscillatory nature of the transient. Resistance is the energy-dissipating element that allows the transient to dampen out and decay to the steady-state condition. Figure 3.1 illustrates an electrical power source feeding a resistive–inductive load (e.g., a motor) via circuit breaker S and transformer T. Figure 3.2 is a steady-state representation of the power circuit, and Figure 3.3 is a transient model of the same circuit. In the transient model, the capacitance across the poles of the circuit breaker, the capacitance

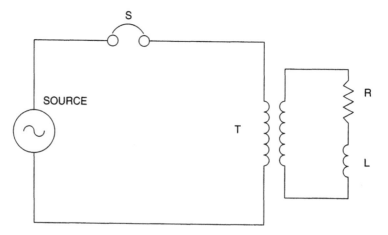

FIGURE 3.1 Power system consisting of source, circuit breaker, transformer, and load.

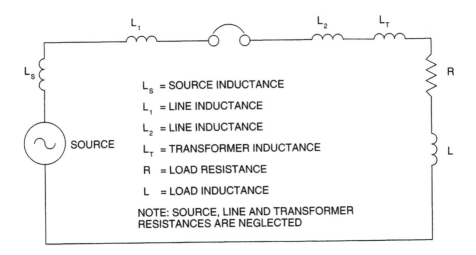

FIGURE 3.2 Low-frequency representation of the circuit shown in Figure 3.1.

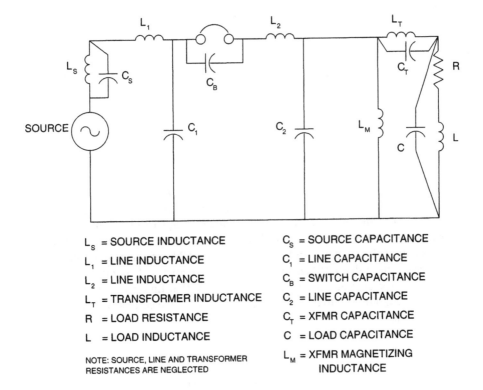

FIGURE 3.3 Transient model of circuit in Figure 3.1.

of the power lines feeding the motor, and the capacitance of the source and the motor windings become significant. Once the transient model is created, some of the elements may be systematically eliminated depending on their magnitude, the transient duration, and the relevance of a specific element to the problem being addressed. This brings up some important points to consider while solving electrical transient-related problems. First, determine the total transient model, then remove elements in the model that are not relevant to the problem at hand. Also, develop a mathematical model of the transient circuit, and then derive a solution for the needed parameter.

3.3 EXAMPLES OF TRANSIENT MODELS AND THEIR RESPONSE

Some examples of transient models and transient response are included in this section. The examples are simple but nonetheless are intended to provide the reader with a better grasp of the basics of electrical transients.

3.3.1 APPLICATION OF DC VOLTAGE TO A CAPACITOR

Figure 3.4 depicts a capacitor to which a DC voltage (V) is suddenly applied. Figure 3.5 represents in graphical form the transient response for the current and voltage across the capacitor. The expressions for V_C and I_C assume that the capacitor has zero initial charge:

$$V_C = V(1 - e^{-t/RC}) \tag{3.1}$$

$$I_C = (V/R)e^{-t/RC} \tag{3.2}$$

where RC is the time constant (T) of the resistance–capacitance circuit and is expressed in seconds. The time constant is the time it would take for an exponentially decaying parameter to reach a value equal to 36.79% of the initial value. This is explained by noting that the parameter would be reduced to a value given by $1/e^1$ or 0.3679 of the initial value. In a time interval equal to two time constants, the parameter will be reduced to $1/e^2$ or 13.5% of the initial value. After five time constants, the parameter will be reduced to 0.67% of the initial value. As time increases, the value $e^{-t/T}$ becomes smaller and smaller and approaches zero. In this example, the current through the capacitor is V/R at time $t = 0$. At time $t = T$, the current will diminish to $0.3679(V/R)$; at time $t = 2T$, the current will be $0.1353(V/R)$, and so on.

In the same example, if the capacitance has an initial voltage of $+V_0$, then the expressions become:

$$V_C = V - (V - V_0)e^{-t/RC} \tag{3.3}$$

$$I_C = [(V - V_0)/R]e^{-t/RC} \tag{3.4}$$

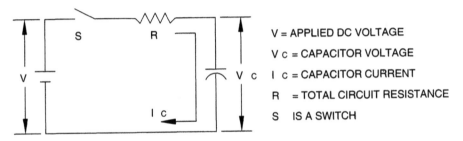

V = APPLIED DC VOLTAGE

V_C = CAPACITOR VOLTAGE

I_C = CAPACITOR CURRENT

R = TOTAL CIRCUIT RESISTANCE

S IS A SWITCH

FIGURE 3.4 Application of DC voltage to an $R–C$ circuit.

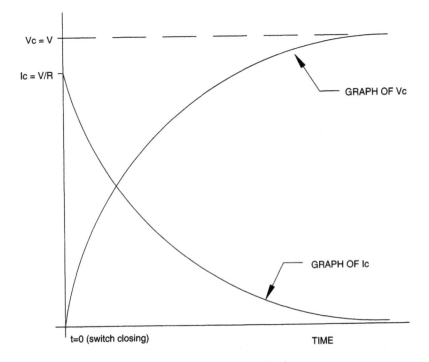

FIGURE 3.5 Capacitor voltage and capacitor current with time.

3.3.2 APPLICATION OF DC VOLTAGE TO AN INDUCTOR

Figure 3.6 shows an inductor with R representing the resistance of the connecting wires and the internal resistance of the inductor. The variation of voltage and the current in the inductor are shown in Figure 3.7 and expressed by:

$$V_L = Ve^{-tR/L} \tag{3.5}$$

$$I_L = (V/R)(1 - e^{-tR/L}) \tag{3.6}$$

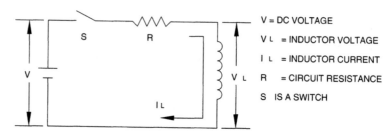

FIGURE 3.6 Application of DC voltage to a R–L circuit.

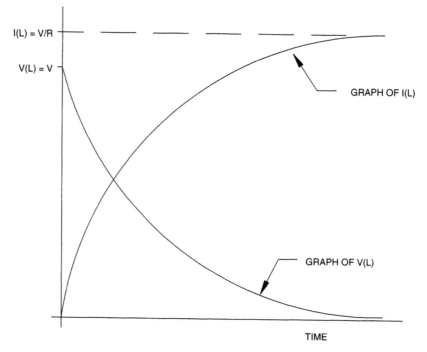

FIGURE 3.7 Inductor voltage and inductor current with time.

At time $t = 0$, the voltage across the inductor is V; in an ideal inductor, current cannot change instantly, so the applied voltage appears across the inductor. At time $t = \infty$, voltage across the inductor is zero. The current through the ideal inductor at time $t = 0$, and at time $t = \infty$ the current through the inductor is equal to V/R.

For an inductive circuit, the time constant T in seconds is equal to L/R, and the expressions for V_L and I_L become:

$$V_L = Ve^{-t/T} \tag{3.7}$$

$$I_L = (V/R)(1 - e^{-t/T}) \tag{3.8}$$

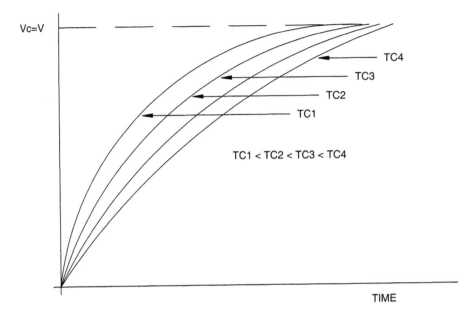

FIGURE 3.8 Variation of V_C with time and with time constant RC.

The significance of the time constant is again as indicated under the discussion for capacitors. In this example, the voltage across the inductor after one time constant will equal 0.3679 V; in two time constants, 0.1353 V; and so on.

The significance of the time constant T in both capacitive and inductive circuits is worth emphasizing. The time constant reflects how quickly a circuit can recover when subjected to transient application of voltage or current. Consider Eq. (3.1), which indicates how voltage across a capacitor would build up when subjected to a sudden application of voltage V. The larger the time constant RC, the slower the rate of voltage increase across the capacitor. If we plot voltage vs. time characteristics for various values of time constant T, the family of graphs will appear as shown in Figure 3.8. In inductive circuits, the time constant indicates how quickly current can build up through an inductor when a switch is closed and also how slowly current will decay when the inductive circuit is opened. The time constant is an important parameter in the transient analysis of power line disturbances.

The L–C combination, whether it is a series or parallel configuration, is an oscillatory circuit, which in the absence of resistance as a damping agent will oscillate indefinitely. Because all electrical circuits have resistance associated with them, the oscillations eventually die out. The frequency of the oscillations is called the natural frequency, f_0. For the L–C circuit:

$$f_0 = 1/2\pi\sqrt{LC} \qquad\qquad (3.9)$$

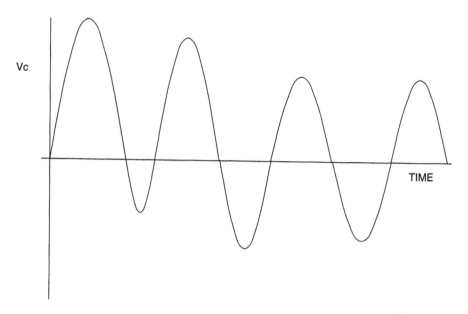

Vc

TIME

FIGURE 3.9 Oscillation of capacitor voltage when *L–C* circuit is closed on a circuit of DC voltage *V.*

In the *L–C* circuit, the voltage across the capacitor might appear as shown in Figure 3.9. The oscillations are described by the Eq. (3.10), which gives the voltage across the capacitance as:

$$V_C = V - (V - V_{CO})\cos\omega_0 t \qquad (3.10)$$

where V is the applied voltage, V_{CO} is the initial voltage across the capacitor, and ω_0 is equal to $2\pi f_0$.

Depending on the value and polarity of V_{CO}, a voltage of three times the applied voltage may be generated across the capacitor. The capacitor also draws a considerable amount of oscillating currents. The oscillations occur at the characteristic frequency, which can be high depending on the value of *L* and *C*. A combination of factors could result in capacitor or inductor failure. Most power systems have some combination of inductance and capacitance present. Capacitance might be that of the power factor correction devices in an electrical system, and inductance might be due to the power transformer feeding the electrical system.

The examples we saw are for *L–C* circuits supplied from a direct current source. What happens when an *L–C* circuit is excited by an alternating current source? Once again, oscillatory response will be present. The oscillatory waveform superimposes on the fundamental waveform until the damping forces sufficiently attenuate the oscillations. At this point, the system returns to normal operation. In a power system characterized by low resistance and high values of *L* and *C*, the effects would be more damaging than if the system were to have high resistance and low *L* and *C* because the natural frequencies are high when the values of *L* and *C* are low. The

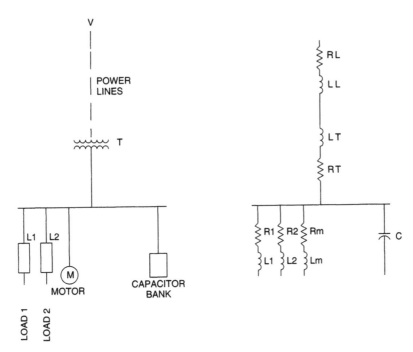

FIGURE 3.10 Lumped parameter representation of power system components.

resistance of the various components that make up the power system is also high at the higher frequencies due to the skin effect, which provides better damping characteristics. In all cases, we are concerned about not only the welfare of the capacitor bank or the transformer but also the impact such oscillations would have on other equipment or devices in the electrical system.

3.4 POWER SYSTEM TRANSIENT MODEL

At power frequencies, electrical systems may be represented by lumped parameters of R, L, and C. Figure 3.10 shows a facility power system fed by 10 miles of power lines from a utility substation where the power is transformed from 12.47 kV to 480 V to supply various loads, including a power factor correction capacitor bank. Reasonable accuracy is obtained by representing the power system components by their predominant electrical characteristics, as shown in Figure 3.10. Such a representation simplifies the calculations at low frequencies. To obtain higher accuracy as the frequency goes up, the constants are divided up and grouped to form the π or T configurations shown in Figure 3.11; the computations get tedious, but more accurate results are obtained. Yet, at high frequencies the power system should be represented by distributed parameters, as shown in Figure 3.12. In Figure 3.12, r, l, and c represent the resistance, inductance, and capacitance, respectively, for the unit distance. The reason for the distributed parameter approach is to produce results that more accurately represent the response of a power system to high-frequency transient phenomena.

ΤΤ REPRESENTATION
OF POWER LINES

T REPRESENTATION
OF POWER LINES

FIGURE 3.11 Representation of power lines at high frequencies where L is the total inductance and C is the total capacitance of the power lines.

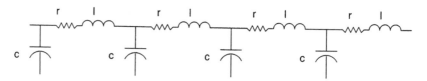

FIGURE 3.12 Distributed constant representation of power lines at high frequencies where c, l, and r are electrical constants for unit distance.

The wavelength of a periodic waveform is given by:

$$\lambda = C/f$$

where C is the velocity of light in vacuum and is equal to 300×10^6 msec or 186,400 miles/sec. For 60-Hz power frequency signals, λ is equal to 3106 miles; for a 1-MHz signal, λ is equal to 393 ft.

All alternating current electrical signals travel on a conducting medium such as overhead power lines or underground cables. When a signal reaches the end of the wiring, it reflects back. Depending on the polarity and the phase angle of the reflected wave, the net amplitude of the composite waveform can have a value between zero and twice the value of the incident wave. Typically, at 1/4 wavelength and odd multiples of 1/4 wavelength, the reflected wave becomes equal in value but opposite in sign to the incident wave. The incident and the reflected waves cancel out, leaving zero net signal. The cable, in essence, acts like a high-impedance circuit. For transient phenomena occurring at high frequencies, however, even comparatively short lengths of wire might be too long to be effective.

Several quantities characterize the behavior of power lines as far as transient response is concerned. One important quantity is the characteristic impedance, expressed as:

$$Z_0 = \sqrt{(L/C)} \qquad (3.11)$$

In a power line that has no losses, the voltage and the current are linked by the characteristic impedance Z_0.

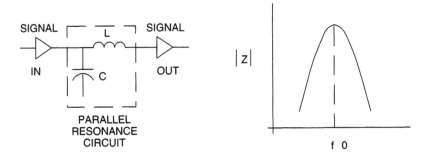

FIGURE 3.13 Parallel resonance circuit and impedance graph indicating highest impedance at the frequency of resonance.

Another important characteristic of power systems is the natural frequency, which allows us to calculate the frequency of a disturbance produced in the *L–C* circuit when it is excited by a voltage or current signal. Why is this important? Transient phenomena are very often oscillatory, and the frequencies encountered are higher than the power frequency. By knowing the circuit constants *L* and *C* and the amplitude of the exciting voltage and current, the response of a transient circuit might be determined with reasonable accuracy. Also, when two circuits or power lines are connected together, the characteristic impedance of the individual circuits determines how much of the transient voltage or current will be reflected back and what portion will be refracted or passed through the junction to the second circuit. This is why, in transient modeling, impedance mismatches should be carefully managed to minimize large voltage or current buildups. The natural frequency is given by:

$$f_0 = 1/2\pi\sqrt{LC} \tag{3.12}$$

Because any electrical signal transmission line has inductance and capacitance associated with it, it also has a natural frequency. The phenomenon of resonance occurs when the capacitive and inductive reactances of the circuit become equal at a given frequency. In transmission line theory, the resonant frequency is referred to as the characteristic frequency. Resonance in a parallel circuit is characterized by high impedance at the resonant frequency, as shown in Figure 3.13. The electrical line or cable has a characteristic resonance frequency that would allow the cable to appear as a large impedance to the flow of current. These typically occur at frequencies corresponding to 1/4 wavelengths. The significance of this becomes apparent when cables are used for carrying high-frequency signals or as ground reference conductors. Conductor lengths for these applications have to be kept short to eliminate operation in the resonance regions; otherwise, significant signal attenuation could result. If the cable is used as a ground reference conductor, the impedance of the cable could render it less than effective.

The velocity of propagation (*v*) indicates how fast a signal may travel in a medium and is given by:

$$v = 1/\sqrt{(\mu\varepsilon)} \tag{3.13}$$

where μ is the permeability of the medium and ε is the dielectric permittivity. For example, in a vacuum, the permeability $= \mu = 4\pi 10^{-7}$ H/m, and the dielectric permittivity $= \varepsilon = 8.85 \times 10^{-12}$ F/m. Therefore,

$$v = 1/\sqrt{(4\pi 10^{-7} \times 8.85 \times 10^{-12})} \cong 300 \times 10^6 \text{ msec} = \text{velocity of light}$$

For other media, such as insulated cables or cables contained in magnetic shields (conduits, etc.), the velocity of propagation will be slower, as these items are no longer characterized by the free air qualities of μ and ε.

The quantities Z_0, f_0, and v are important for examining transient phenomenon because high-speed, high-frequency events can travel through a conductive path (wire) or may be coupled to adjacent circuits by propagation through a dielectric medium. How effective the path is at coupling the transient depends on these factors.

3.5 TYPES AND CAUSES OF TRANSIENTS

Transients are disturbances that occur for a very short duration (less than a cycle), and the electrical circuit is quickly restored to original operation provided no damage has occurred due to the transient. An electrical transient is a cause-and-effect phenomenon. For transients to occur, there must be a cause. While they may be many, this section will look at some of the more common causes of transients:

- Atmospheric phenomena (lightning, solar flares, geomagnetic disturbances)
- Switching loads on or off
- Interruption of fault currents
- Switching of power lines
- Switching of capacitor banks

3.5.1 ATMOSPHERIC CAUSES

Over potential surge due to lightning discharge is the most common natural cause of electrical equipment failure. The phenomenon of lighting strike can be described as follows. A negative charge builds up on a cloud, as indicated in Figure 3.14. A corresponding positive charge can build up on the surface of the earth. A voltage difference of hundreds of millions of volts can exist between the cloud and the earth due to the opposing charges. When the voltage exceeds the breakdown potential of air (about 3×10^6 V/m or 75 kV/inch), a lightning flash occurs. The exact physics of the lightning phenomenon will not be discussed here, as it is sufficient to know that a lightning strike can typically produce a voltage rise in about 1 or 2 μsec that can decline to a value of 50% of the peak voltage in approximately 50 to 100 μsec. A typical lightning impulse wave might appear as shown in Figure 3.15. A common misconception is that a direct lightning strike is needed to produce destructive overvoltages. In fact, it is rare that a failure in an electrical system is due to a direct lightning strike. More often, the electrical and magnetic fields caused by indirect lightning discharge induce voltages in the power lines that result in device failures. Also, lightning discharge current flowing through the earth creates a potential dif-

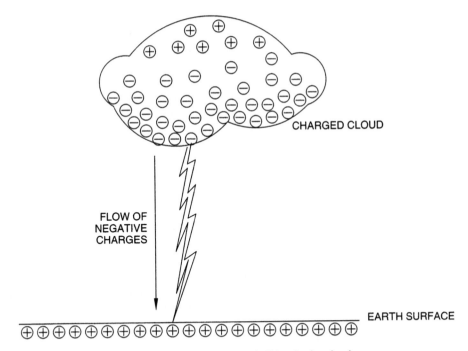

FIGURE 3.14 Lightning discharge due to charge buildup in the clouds.

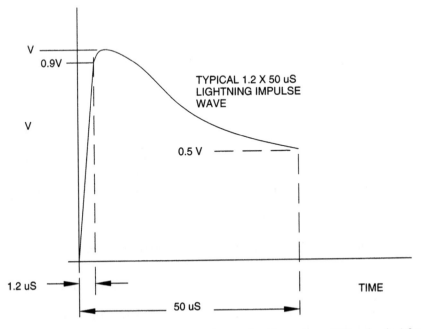

FIGURE 3.15 Lightning impulse waveform characterized by a rise to 90% value in 1.2 μs and a fall to 50% value in 50 μs.

ference between the power lines and ground and in extreme cases causes equipment failure.

Isolation transformers provide limited protection from lightning strikes. Because lightning is a short-duration, high-frequency phenomenon, a portion of the lightning energy will couple directly from the primary winding to the secondary winding of the transformer through the interwinding capacitance. This is why equipment supplied from the low-voltage winding of a transformer that is exposed to lightning energy is also at risk. The amount of voltage that will be coupled through the transformer will depend on the transformer interwinding capacitance itself. The higher the capacitance, the higher the transient energy coupled to the secondary. Transformers provided with a grounded shield between the primary and the secondary windings provide better protection against lightning energy present at the transformer primary winding.

Lightning arresters, when properly applied, can provide protection against lightning-induced low voltages. Arresters have a well-defined conduction voltage below which they are ineffective. This voltage depends on the rating of the arrester itself. For optimum protection, the arrester voltage should be matched to the lightning impulse withstand of the equipment being protected. Table 3.1 provides typical voltage discharge characteristics of arresters for various voltage classes.

3.5.2 SWITCHING LOADS ON OR OFF

Switching normal loads in a facility can produce transients. The majority of plant loads draw large amounts of current when initially turned on. Transformers draw

TABLE 3.1
Typical Surge Arrester Protective Characteristics

Arrester Rating (kV rms)	Maximum Continuous Operating Voltage[a] (kV rms)	1-sec Temporary Overvoltage (kV rms)	Maximum Front of Wave Protective Level[b] (kV crest)
3	2.55	4.3	10.4
6	5.1	8.6	20.7
9	7.65	12.9	31.1
12	10.2	17.2	41.5
15	12.7	21.4	51.8
18	15.3	25.8	62.2
21	17.0	28.7	72.6
24	19.5	32.9	82.9

Note: For proper protection, the impulse level of all protected equipment must be greater than the front of wave protective level by a margin of 25% or greater.

[a] Maximum continuous operating voltage is the maximum voltage at which the arrester may be operated continuously.

[b] Maximum front of wave protective level is the kilovolt level at which the arrester clamps the front of the impulse waveform.

inrush currents that range between 10 and 15 times their normal full-load current. This current lasts between 5 and 10 cycles. Alternating current motors draw starting currents that vary between 500 and 600% of the normal full-load running current. Fluorescent lights draw inrush currents when first turned on. Large current drawn through the impedance of the power system sets up transient voltages that affect electrical components sensitive to sags, subcycle oscillations, or voltage notch. There are instances when conditions are such that harmonic frequency currents in the inrush current interact with the power system inductance and capacitance and cause resonance conditions to develop. During resonance, substantial overvoltages and overcurrents might develop. In the strict sense, these are not subcycle events and therefore may not be classified as transients, but their effects are nonetheless very detrimental.

Large inrush currents drawn by certain loads produce other negative effects. Consider a conductor carrying a large current. The magnetic field due to the surge current could induce large potentials in adjacent signal or data cables by inductive coupling. This is why it is preferable to keep signal or data cables physically distant from power cables. Data and signal wires that run near power cables should be contained in metal conduits made of steel. Steel, due to its magnetic properties, is a better shield at low frequencies than nonferrous metals such as aluminum or copper. Nonferrous metals make better shields at high frequencies.

When discussing inductive coupling due to transient current, the loop area of the susceptible circuit should not be overlooked (see Figure 3.16), as the larger the area of the loop, the higher the noise voltage induced in the susceptible circuit. In Figure 3.16, the voltage induced in circuit 2 depends on the magnetic field linking the circuit; the larger the loop area, the larger the flux linkage and, therefore, the higher the noise voltage induced in circuit 2. Twisted pairs of wires minimize the loop area and reduce noise voltage pickup, thus signal and data circuits for sensitive, low-level signal applications installed in close proximity to power wires should use twisted sets of wires to reduce noise coupling.

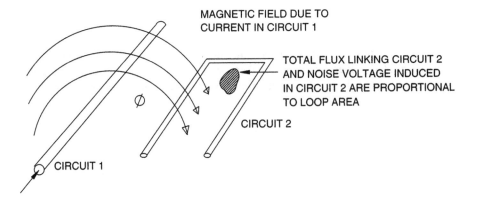

MAGNETIC FIELD DUE TO CURRENT IN CIRCUIT 1

TOTAL FLUX LINKING CIRCUIT 2 AND NOISE VOLTAGE INDUCED IN CIRCUIT 2 ARE PROPORTIONAL TO LOOP AREA

CIRCUIT 2

CIRCUIT 1

FIGURE 3.16 Voltage induced in circuit 2 due to current in circuit 1. The voltage depends on the loop area of circuit 2 and proximity between the circuits.

So far we have examined the effects of switching power to loads during normal operation. Switching power off also generates transients due to the fact that all devices carrying electrical current have inductances (L) associated with them. The current flowing in an inductive device cannot abruptly change when the circuit is opened. The voltage produced in an electrical device due to a sudden change of current is given by:

$$e = L \times di/dt$$

where di/dt is the rate change of current and L is the inductance associated with the device.

For example, if 50 A of current flowing through a coil of inductance $L = 20$ mH drops to zero in 2 msec, then the voltage generated across the coil is given by:

$$E = (20 \times 10^{-3} \times 50)/(2 \times 10^{-3}) = 500 \text{ V}$$

It is easy to see that substantial voltages can be developed while switching inductive devices off. The transient voltage produced can easily couple to other circuits via stray capacitance between the inductive device and the susceptible circuit. This voltage can appear as noise for the second circuit. The closer the two circuits are spaced, the higher the amount of noise that is coupled. Voltages as high as 2000 to 5000 V are known to be generated when large inductive currents are interrupted. In low-voltage power and signal circuits, this can easily cause damage.

3.5.3 Interruption of Fault Circuits

During fault conditions, large currents are generated in an electrical system. The fault currents are interrupted by overcurrent devices such as circuit breakers or fuses. Figure 3.17 shows a simplified electrical circuit where an electrical fault is cleared by a circuit breaker. C represents the capacitance of the electrical system up to the point where the overcurrent device is present. Interruption of the fault current generates overvoltage impulse in the electrical system, and the magnitude of the voltage depends on the amount of fault current and the speed with which the fault is interrupted. Older air circuit breakers with slower speed of interruption produce lower impulse voltages than vacuum or SF_6 breakers, which operate at much faster

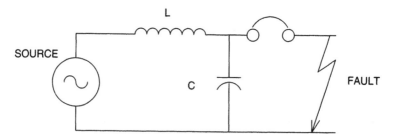

FIGURE 3.17 Electrical fault at the output side of a circuit breaker.

speeds during a fault. While using vacuum or SF_6 technology to clear a fault quickly and thereby limit damage to equipment is an important advantage, a price is paid by the generation of higher level voltage transients. Once the fault is interrupted, the generated voltage impulse can interact with the inductance and capacitance of the electrical system and produce oscillation at a frequency much higher than the fundamental frequency. The oscillations are slowly damped out by the resistance associated with the system. The response of the system might appear somewhat like what is shown in Figure 3.18. The voltage can build up to levels equal to twice the peak value of the voltage waveform. The overvoltage and associated oscillations are harmful to electrical devices.

A very serious case of overvoltages and oscillations occurs when the overcurrent device is supplied from overhead power lines that connect to long lengths of underground cables. Underground cables have substantial capacitance to ground. The combination of the inductance due to overhead lines and capacitance due to underground cables could generate high levels of overvoltage and prolonged oscillations at low frequencies. Such transients are very damaging to transformers, cables, and motors supplied from the lines. In extreme cases, voltages as high as three to four times the AC peak voltage may be generated.

3.5.4 Capacitor Bank Switching

One of the more common causes of electrical transients is switching of capacitor banks in power systems. Electrical utilities switch capacitor banks during peak load hours to offset the lagging kVAR demand of the load. The leading kVARs drawn

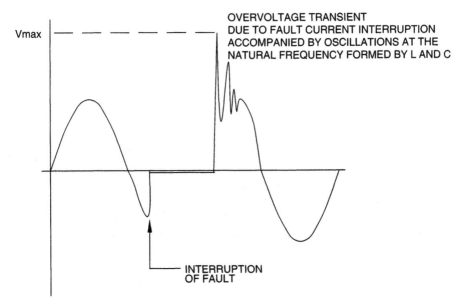

FIGURE 3.18 Overvoltage transient due to fault current interruption shown in Figure 3.17.

by the capacitor banks offset the lagging kVAR demand of the load, reducing the net kVA load on the circuit. Switching of capacitor banks is accompanied by a surge of current which is initially limited by the characteristic impedance of the power system and resistance of the line. A sharp reduction in the voltage is followed by a voltage rise, which decays by oscillation at a frequency determined by the inductance and capacitance of the circuit. Several cases of power system component failures and malfunctions due to capacitor bank switching operations have been seen by the author. Typically, the voltage rise due to capacitor switching operation can attain values 1.5 to 2 times the nominal voltage. Power equipment can withstand only a limited number of exposures to such rises in voltage magnitude. With time, the insulation systems of such devices weaken, and a point is reached when the devices can fail. In one particular instance, two power distribution transformers failed at the same time; the cause was traced to large capacitor bank switching operations by the utility at a substation located adjacent to the affected facility.

Adjustable speed drives (ASDs) and solid-state motor controllers are quite sensitive to voltage rises resulting from capacitor bank switching operations. The ASD might shut down the motor due to voltage on the system rising beyond the maximum tolerance. In some cases, capacitor switching causes the voltage waveform to undergo oscillations and produce stray crossings of the time axis. This is unacceptable for devices that require the precise number of zero time crossings for proper performance.

Example: A 2000-kVAR, 13.8-kV, Y-connected capacitor bank is connected at the end of a 25-mile transmission line with an inductive reactance of 0.5 Ω per mile. Find the natural frequency of the current that would be drawn during turn on:

Total inductive reactance = X_L = 25 × 0.5 = 12.5 Ω
Inductance (L) = 12.5/120π = 0.033 H
Current through the capacitor bank (I_C) = 83.7 A
Capacitive reactance (X_C) = 7967/83.7 = 95.18 Ω
Capacitance (C) = 27.9 µF
Characteristic frequency (f_0) = 1/(2$X\pi$ $\sqrt{0.033 \times 0.0000279}$) $\cong$ 166 Hz

The current drawn from the source will have a frequency of 166 Hz which will decay as determined by the power system resistance. Due to impedance drops associated with the currents, the voltage waveform would experience similar oscillations prior to settling down to nominal levels. The series resonance circuit formed by the system inductance and the capacitance could produce a voltage rise in the electrical system. Depending on the severity of the voltage rise and the time to decay, equipment damage can result, especially if the switching operations are frequent. The condition is made worse if a second capacitor bank is brought online. The natural frequencies associated with this action are higher and so is the time to decay. Considerable energy is exchanged between the two capacitors before steady-state operation is attained.

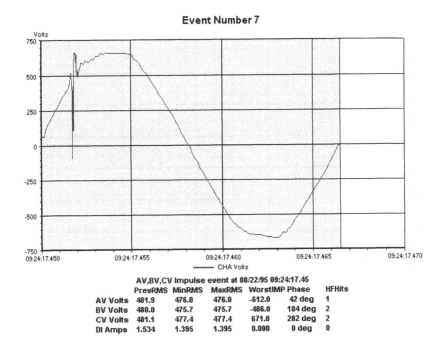

FIGURE 3.19 Transient due to motor starting. The motor had an input capacitor for power factor correction, and the motor and capacitor were turned on simultaneously.

3.6 EXAMPLES OF TRANSIENT WAVEFORMS

3.6.1 MOTOR START TRANSIENT

Figure 3.19 shows a transient produced when a 50-hp induction motor with integral power factor correction was started across the line. The notch in the voltage waveform at the instant of starting was produced by the presence of the capacitor. The quick voltage recovery was followed by ringing characteristics. The transient lasted less than half of a cycle, but it was sufficient to affect the operation of large chillers located nearby which contained solid-state starters with sensitive voltage-sensing circuitry. Because of the severity of the motor-starting transient, the chillers started to shut down. In a situation such as this, it is often prudent to apply correction to the sensitive circuitry rather than try to eliminate the problem itself or apply correction to the power system as a whole.

3.6.2 POWER FACTOR CORRECTION CAPACITOR SWITCHING TRANSIENT

Figure 3.20 shows the transient voltage response at the main electrical switchboard for a commercial building due to capacitor bank switching by the utility. A moderate

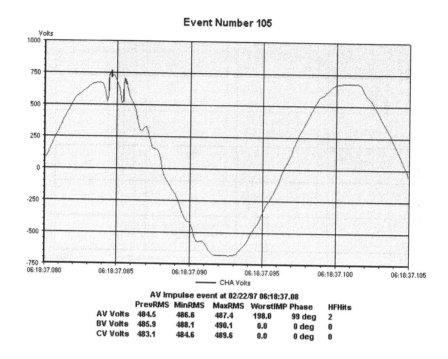

FIGURE 3.20 Transient due to capacitor bank switching by the utility. The waveform was recorded at the main electrical switchboard for a commercial building.

rise in system voltage is followed by ringing at the characteristic frequency of the utility source inductive reactance and capacitance due to the power factor correction equipment.

3.6.3 MEDIUM VOLTAGE CAPACITOR BANK SWITCHING TRANSIENT

The transient shown in Figure 3.21 was the result of a 5-MVAR capacitor bank switching at an industrial facility. The facility suffered from poor power factor due to plant loads which necessitated connection of the capacitor bank. The initial rise in voltage reached peak amplitude equal to 160% of the system nominal peak voltage. This practice had been going on in this facility for approximately 5 years before failures were observed in underground cables and a power transformer. At this point, the electrical system in the facility was monitored for transient voltages by installing power quality analyzers at select locations. Once the nature of the transients and their cause were determined, corrective steps were taken to retrofit the capacitor bank with pre-insertion resistors, which helped to attenuate the amplitude of the impulse. Also, all replacement equipment at the facility was specified to be of a higher basic insulation level (BIL) than the minimum specified in standards.

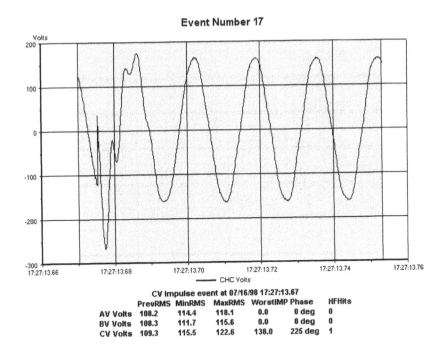

FIGURE 3.21 Voltage waveform at a 12.47-kV power system during switching in of a 5-MVAR capacitor bank. The voltage-to-transformer ratio is 60:1.

3.6.4 Voltage Notch Due to Uninterruptible Power Source Unit

Typically, we associate voltage notches with adjustable speed drives. Voltage notches are also common with the outputs of uninterruptible power source (UPS) units due to power electronic switching circuitry associated with the UPS units. Unless provided with wave-shaping and filtering circuitry, the output of the UPS can contain substantial notches. Figure 3.22 shows the output waveform of a UPS unit supplying 480-V output to a computer center at a financial institution. If the notch levels become excessive, problems can arise in the operation of sensitive communication or data-processing loads. The voltage notch phenomenon is a repetitive event. Even though we defined transients as subcycle events, the repetitive notching shown is included in this section for the sake of completeness.

3.6.5 Neutral Voltage Swing

The event shown in Figure 3.23 was observed in a computer laboratory at a university. Normally, neutral voltage should be within 0.5 V with respect to the ground. This is because in a four-wire power distribution system the neutral of the power source

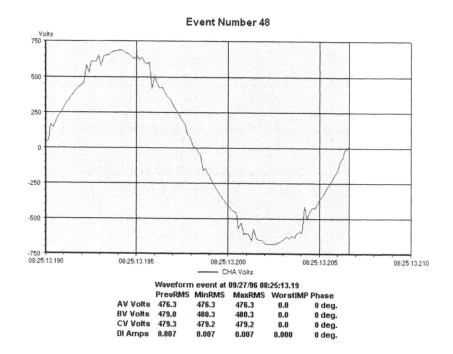

FIGURE 3.22 Voltage notches produced at the output of an uninterruptible power source (UPS) unit.

is connected to the ground at the source. This tends to hold the neutral potential close to the ground. In a typical building, neutral-to-ground voltages become higher as we move away from the source feeding the facility. In some instances, depending on the loads and the distance between the source and the load, neutral-to-ground voltage can measure 2 to 3 V. The case illustrated by Figure 3.23 is an extreme one where neutral-to-ground voltages reached levels higher than 10 V. The computer laboratory experienced many problems with the computers locking up, in some instances during critical times when tests were being administered to the students.

3.6.6 Sudden Application of Voltage

Fast rise time transients are produced when voltage is suddenly applied to a load. Figure 3.24 shows an example of 480 V being applied to the primary of a power distribution transformer. Typical waveform characteristics include fast rise time and ringing due to the inductance and capacitance of the load circuitry. Normally, power system should ride through such occurrences, but, if the load circuit includes capacitor banks or power supplies with capacitors, large inrush currents may be produced with possible overcurrent protection operation. In situations where the capacitors have an initial charge present, some overvoltage events may be produced.

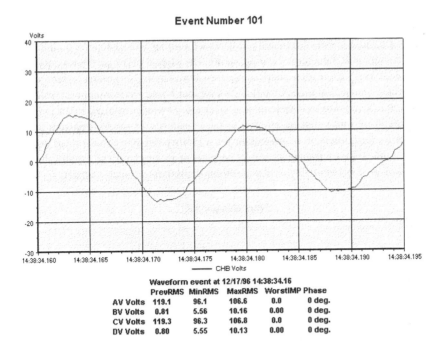

FIGURE 3.23 Large neutral voltage swings responsible for problems in a university computer laboratory.

3.6.7 SELF-PRODUCED TRANSIENTS

Some machines by nature generate transients that can affect machine operation if they contain sensitive circuits. Figure 3.25 shows a waveform produced by a food-processing machine. The transients occurred several times each day as the machine automatically turned on and off. Severe transients caused the machine to go into a lockout mode which required the operator to manually reset the machine. The problem was fixed by providing a dedicated circuit of low impedance for the machine. This reduced the transient to levels that could be lived with.

3.7 CONCLUSIONS

A power system during normal operation is in a steady-state condition even though some voltage fluctuations may be present as the result of facility or utility switching operations. The voltage stays within tolerances that would normally be expected. All electrical and electronic devices are designed to function within these tolerances. Some level of degradation of the useful life of the equipment is to be expected even though the operating voltages are within tolerances. Transients are abnormal, short-duration power system events that have a specific cause behind them, such as switching operations or electrical faults or even nature-induced events. Very few devices are designed specifically with transients in mind, and most devices can handle a limited number of transients. The exact number would depend on the nature of the transient and the age of the equipment. The effects of transients on equipment are cumulative, with every succeeding transient having a greater effect on the equipment. Electrical devices installed in a typical home environment are relatively safe as far as exposure to transients is concerned. Devices installed in an industrial environment are more susceptible due to the possibility of severe transient activity in such an environment. It is important for a facility designer or operator not only to know what type of transients might be present in an electrical system, but also to be aware of the sensitivity of the installed equipment to such transients.

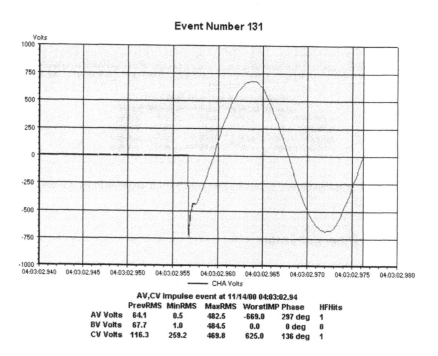

Event Number 131

AV,CV Impulse event at 11/14/00 04:03:02.94					
	PrevRMS	MinRMS	MaxRMS	WorstIMP	Phase	HFHits
AV Volts	64.1	0.5	482.5	-669.0	297 deg	1
BV Volts	67.7	1.0	484.5	0.0	0 deg	0
CV Volts	116.3	259.2	469.8	625.0	136 deg	1

FIGURE 3.24 Fast rise transient generated when a 480-V feeder was energized. The transient produced ringing due to system inductance and capacitance.

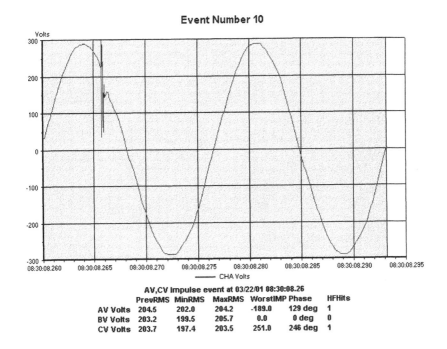

Event Number 10

AV,CV Impulse event at 03/22/01 08:30:08.26

	PrevRMS	MinRMS	MaxRMS	WorstIMP	Phase	HFHits
AV Volts	204.5	202.0	204.2	-189.0	129 deg	1
BV Volts	203.2	199.5	205.7	0.0	0 deg	0
CV Volts	203.7	197.4	203.5	251.0	246 deg	1

FIGURE 3.25 Transient produced by a machine itself. This event was recorded at the supply lines to a food-processing machine. At the start of each operation, the machine generated transients, which, when severe enough, shut down the machine.

4 Harmonics

4.1 DEFINITION OF HARMONICS

Webster's New World Dictionary defines *harmonics* as pure tones making up a composite tone in music. A pure tone is a musical sound of a single frequency, and a combination of many pure tones makes up a composite sound. Sound waves are electromagnetic waves traveling through space as a periodic function of time. Can the principle behind pure music tones apply to other functions or quantities that are time dependent? In the early 1800s, French mathematician, Jean Baptiste Fourier formulated that a periodic nonsinusoidal function of a fundamental frequency f may be expressed as the sum of sinusoidal functions of frequencies which are multiples of the fundamental frequency. In our discussions here, we are mainly concerned with periodic functions of voltage and current due to their importance in the field of power quality. In other applications, the periodic function might refer to radio-frequency transmission, heat flow through a medium, vibrations of a mechanical structure, or the motions of a pendulum in a clock.

A sinusoidal voltage or current function that is dependent on time t may be represented by the following expressions:

$$\text{Voltage function, } v(t) = V \sin(\omega t) \tag{4.1}$$

$$\text{Current function, } i(t) = I \sin(\omega t \pm \varnothing) \tag{4.2}$$

where $\omega = 2 \times \pi \times f$ is known as the angular velocity of the periodic waveform and $\varnothing$ is the difference in phase angle between the voltage and the current waveforms referred to as a common axis. The sign of phase angle $\varnothing$ is positive if the current leads the voltage and negative if the current lags the voltage. Figure 4.1 contains voltage and current waveforms expressed by Eqs. (4.1) and (4.2) and which by definition are pure sinusoids.

For the periodic nonsinusoidal waveform shown in Figure 4.2, the simplified Fourier expression states:

$$v(t) = V_0 + V_1 \sin(\omega t) + V_2 \sin(2\omega t) + V_3 \sin(3\omega t) + \ldots + V_n \sin(n\omega t) + V_{n+1} \sin((n+1)\omega t) + \ldots \tag{4.3}$$

The Fourier expression is an infinite series. In this equation, V_0 represents the constant or the DC component of the waveform. $V_1, V_2, V_3, \ldots, V_n$ are the peak values of the successive terms of the expression. The terms are known as the harmonics of the periodic waveform. The fundamental (or first harmonic) frequency has a

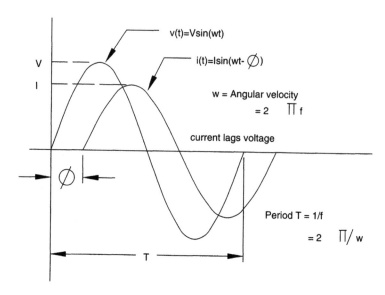

FIGURE 4.1 Sinusoidal voltage and current functions of time (t). Lagging functions are indicated by negative phase angle and leading functions by positive phase angle.

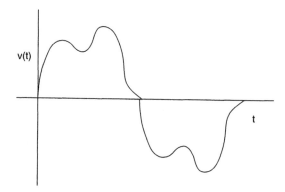

FIGURE 4.2 Nonsinusoidal voltage waveform Fourier series. The Fourier series allows expression of nonsinusoidal periodic waveforms in terms of sinusoidal harmonic frequency components.

frequency of f, the second harmonic has a frequency of $2 \times f$, the third harmonic has a frequency of $3 \times f$, and the nth harmonic has a frequency of $n \times f$. If the fundamental frequency is 60 Hz (as in the U.S.), the second harmonic frequency is 120 Hz, and the third harmonic frequency is 180 Hz.

The significance of harmonic frequencies can be seen in Figure 4.3. The second harmonic undergoes two complete cycles during one cycle of the fundamental frequency, and the third harmonic traverses three complete cycles during one cycle of the fundamental frequency. V_1, V_2, and V_3 are the peak values of the harmonic components that comprise the composite waveform, which also has a frequency of f.

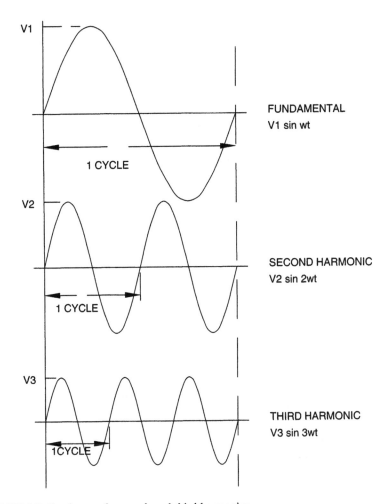

FIGURE 4.3 Fundamental, second, and third harmonics.

The ability to express a nonsinusoidal waveform as a sum of sinusoidal waves allows us to use the more common mathematical expressions and formulas to solve power system problems. In order to find the effect of a nonsinusoidal voltage or current on a piece of equipment, we only need to determine the effect of the individual harmonics and then vectorially sum the results to derive the net effect. Figure 4.4 illustrates how individual harmonics that are sinusoidal can be added to form a nonsinusoidal waveform.

The Fourier expression in Eq. (4.3) has been simplified to clarify the concept behind harmonic frequency components in a nonlinear periodic function. For the purist, the following more precise expression is offered. For a periodic voltage wave with fundamental frequency of $\omega = 2\pi f$,

$$v(t) = V_0 + \sum (a_k \cos k\omega t + b_k \sin k\omega t) \text{ (for } k = 1 \text{ to } \infty) \tag{4.4}$$

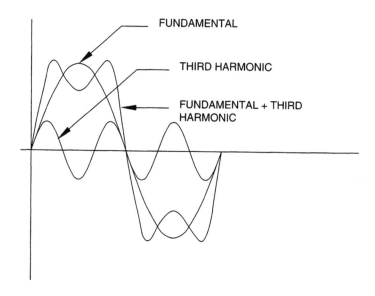

FIGURE 4.4 Creation of nonlinear waveform by adding the fundamental and third harmonic frequency waveforms.

where a_k and b_k are the coefficients of the individual harmonic terms or components. Under certain conditions, the cosine or sine terms can vanish, giving us a simpler expression. If the function is an even function, meaning $f(-t) = f(t)$, then the sine terms vanish from the expression. If the function is odd, with $f(-t) = -f(t)$, then the cosine terms disappear. For our analysis, we will use the simplified expression involving sine terms only. It should be noted that having both sine and cosine terms affects only the displacement angle of the harmonic components and the shape of the nonlinear wave and does not alter the principle behind application of the Fourier series.

The coefficients of the harmonic terms of a function $f(t)$ contained in Eq. (4.4) are determined by:

$$a_k = \frac{1}{\pi} \int_{-\pi}^{+\pi} f(t).\cos kt.dt, \ (k = 1,2,3, \ ..., \ n) \tag{4.5}$$

$$b_k = \frac{1}{\pi} \int_{-\pi}^{+\pi} f(t).\sin kt.dt, \ (k = 1,2,3, \ ..., \ n) \tag{4.6}$$

The coefficients represent the peak values of the individual harmonic frequency terms of the nonlinear periodic function represented by $f(t)$. It is not the intent of this book to explore the intricacies of the Fourier series. Several books in mathematics are available for the reader who wants to develop a deeper understanding of this very essential tool for solving power quality problems related to harmonics.

4.2 HARMONIC NUMBER (*h*)

Harmonic number (*h*) refers to the individual frequency elements that comprise a composite waveform. For example, $h = 5$ refers to the fifth harmonic component with a frequency equal to five times the fundamental frequency. If the fundamental frequency is 60 Hz, then the fifth harmonic frequency is 5×60, or 300 Hz. The harmonic number 6 is a component with a frequency of 360 Hz. Dealing with harmonic numbers and not with harmonic frequencies is done for two reasons. The fundamental frequency varies among individual countries and applications. The fundamental frequency in the U.S. is 60 Hz, whereas in Europe and many Asian countries it is 50 Hz. Also, some applications use frequencies other than 50 or 60 Hz; for example, 400 Hz is a common frequency in the aerospace industry, while some AC systems for electric traction use 25 Hz as the frequency. The inverter part of an AC adjustable speed drive can operate at any frequency between zero and its full rated maximum frequency, and the fundamental frequency then becomes the frequency at which the motor is operating. The use of harmonic numbers allows us to simplify how we express harmonics. The second reason for using harmonic numbers is the simplification realized in performing mathematical operations involving harmonics.

4.3 ODD AND EVEN ORDER HARMONICS

As their names imply, odd harmonics have odd numbers (e.g., 3, 5, 7, 9, 11), and even harmonics have even numbers (e.g., 2, 4, 6, 8, 10). Harmonic number 1 is assigned to the fundamental frequency component of the periodic wave. Harmonic number 0 represents the constant or DC component of the waveform. The DC component is the net difference between the positive and negative halves of one complete waveform cycle. Figure 4.5 shows a periodic waveform with net DC content. The DC component of a waveform has undesirable effects, particularly on transformers, due to the phenomenon of core saturation. Saturation of the core is caused by operating the core at magnetic field levels above the knee of the magnetization curve. Transformers are designed to operate below the knee portion of the curve. When DC voltages or currents are applied to the transformer winding, large DC magnetic fields are set up in the transformer core. The sum of the AC and the DC magnetic fields can shift the transformer operation into regions past the knee of the saturation curve. Operation in the saturation region places large excitation power requirements on the power system. The transformer losses are substantially increased, causing excessive temperature rise. Core vibration becomes more pronounced as a result of operation in the saturation region.

We usually look at harmonics as integers, but some applications produce harmonic voltages and currents that are not integers. Electric arc furnaces are examples of loads that generate non-integer harmonics. Arc welders can also generate non-integer harmonics. In both cases, once the arc stabilizes, the non-integer harmonics mostly disappear, leaving only the integer harmonics.

The majority of nonlinear loads produce harmonics that are odd multiples of the fundamental frequency. Certain conditions need to exist for production of even

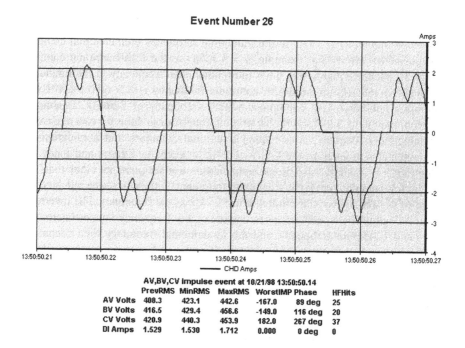

FIGURE 4.5 Current waveform with DC component (scale, 1 A = 200 A). This waveform has a net negative DC component as indicated by the larger area of the negative half compared to the positive half of each cycle.

harmonics. Uneven current draw between the positive and negative halves of one cycle of operation can generate even harmonics. The uneven operation may be due to the nature of the application or could indicate problems with the load circuitry. Transformer magnetizing currents contain appreciable levels of even harmonic components and so do arc furnaces during startup.

Subharmonics have frequencies below the fundamental frequency and are rare in power systems. When subharmonics are present, the underlying cause is resonance between the harmonic currents or voltages with the power system capacitance and inductance. Subharmonics may be generated when a system is highly inductive (such as an arc furnace during startup) or if the power system also contains large capacitor banks for power factor correction or filtering. Such conditions produce slow oscillations that are relatively undamped, resulting in voltage sags and light flicker.

4.4 HARMONIC PHASE ROTATION AND PHASE ANGLE RELATIONSHIP

So far we have treated harmonics as stand-alone entities working to produce waveform distortion in AC voltages and currents. This approach is valid if we are looking at single-phase voltages or currents; however, in a three-phase power system, the harmonics of one phase have a rotational and phase angle relationship with the

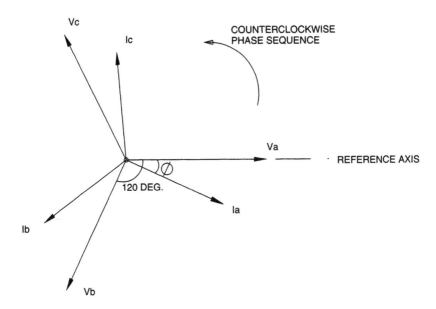

FIGURE 4.6 Balanced three-phase power system. Phase sequence refers to the order in which phasors move past a reference axis. The positive phase sequence is assigned a counterclockwise rotation.

harmonics of the other phases. In power system studies involving harmonics, this relationship is important.

In a balanced three-phase electrical system, the voltages and currents have a positional relationship as shown in Figure 4.6. The three voltages are 120° apart and so are the three currents. The normal phase rotation or sequence is a–b–c, which is counterclockwise and designated as the positive-phase sequence in this book. For harmonic analyses, these relationships are still applicable, but the fundamental components of voltages and currents are used as reference. All other harmonics use the fundamental frequency as the reference. The fundamental frequencies have a positive-phase sequence. The angle between the fundamental voltage and the fundamental current is the displacement power factor angle, as defined in Chapter 1.

So how do the harmonics fit into this space–time picture? For a clearer understanding, let us look only at the current harmonic phasors. We can further simplify the picture by limiting the discussion to odd harmonics only, which under normal and balanced conditions are the most prevalent. The following relationships are true for the fundamental frequency current components in a three-phase power system:

$$i_{a1} = I_{a1} \sin \omega t \tag{4.7}$$

$$i_{b1} = I_{b1} \sin (\omega t - 120°) \tag{4.8}$$

$$i_{c1} = I_{c1} \sin (\omega t - 240°) \tag{4.9}$$

The negative displacement angles indicate that the fundamental phasors i_{b1} and i_{c1} trail the i_{a1} phasor by the indicated angle. Figure 4.7a shows the fundamental current phasors.

The expressions for the third harmonic currents are:

$$i_{a3} = I_{a3} \sin 3\omega t \qquad (4.10)$$

$$i_{b3} = I_{b3} \sin 3(\omega t-120°) = I_{b3} \sin (3\omega t-360°) = I_{b3} \sin 3\omega t \qquad (4.11)$$

$$i_{c3} = I_{c3} \sin 3(\omega t-240°) = I_{c3} \sin (3\omega t-720°) = I_{c3} \sin 3\omega t \qquad (4.12)$$

The expressions for the third harmonics show that they are in phase and have zero displacement angle between them. Figure 4.7b shows the third harmonic phasors. The third harmonic currents are known as zero sequence harmonics due to the zero displacement angle between the three phasors.

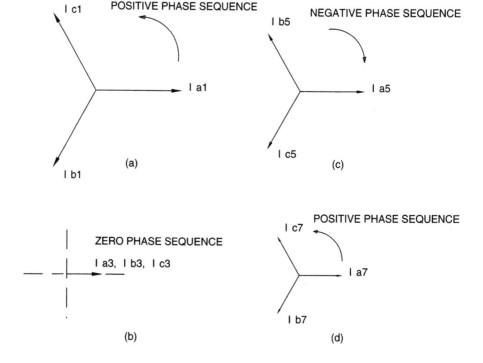

FIGURE 4.7 (a) Fundamental phasors. (b) Third harmonic phasors. (c) Fifth harmonic phasors. (d) Seventh harmonic phasors.

The expressions for the fifth harmonic currents are:

$$i_{a5} = I_{a5} \sin 5\omega t \qquad (4.13)$$

$$i_{b5} = I_{b5} \sin 5(\omega t-120°) = I_{b5} \sin(5\omega t-600°) = I_{b5} \sin(5\omega t-240°) \qquad (4.14)$$

$$i_{c5} = I_{c5} \sin 5(\omega t-240°) = I_{c5} \sin(5\omega t-1200°) = I_{c5} \sin(5\omega t-120°) \qquad (4.15)$$

Figure 4.7c shows the fifth harmonic phasors. Note that the phase sequence of the fifth harmonic currents is clockwise and opposite to that of the fundamental. The fifth harmonics are negative sequence harmonics.

Similarly the expressions for the seventh harmonic currents are:

$$i_{a7} = I_{a7} \sin 7\omega t \qquad (4.16)$$

$$i_{b7} = I_{b7} \sin 7(\omega t-120°) = I_{b7} \sin(7\omega t-840°) = I_{b7} \sin(7\omega t-120°) \qquad (4.17)$$

$$i_{c7} = I_{c7} \sin 7(\omega t-240°) = I_{c7} \sin(7\omega t-1680°) = I_{c7} \sin(7\omega t-240°) \qquad (4.18)$$

Figure 4.7d shows the seventh harmonic current phasors. The seventh harmonics have the same phase sequence as the fundamental and are positive sequence harmonics. So far, we have not included even harmonics in the discussion; doing so is left to the reader as an exercise. Table 4.1 categorizes the harmonics in terms of their respective sequence orders.

TABLE 4.1
Harmonic Order vs. Phase Sequence

Harmonic Order	Sequence
1, 4, 7, 10, 13, 16, 19	Positive
2, 5, 8, 11, 14, 17, 20	Negative
3, 6, 9, 12, 15, 18, 21	Zero

The expressions shown so far for harmonics have zero phase shifts with respect to the fundamental. It is not uncommon for the harmonics to have a phase-angle shift with respect to the fundamental. Figure 4.8 depicts a fifth harmonic current waveform with and without phase shift from the fundamental. Expressions for the fifth harmonics with a phase-shift angle of θ degrees are:

$$i_{a5} = I_{a5} \sin 5(\omega t-\theta) \qquad (4.19)$$

$$i_{b5} = I_{b5} \sin 5(\omega t-120°-\theta) \qquad (4.20)$$

$$i_{c5} = I_{c5} \sin 5(\omega t-240°-\theta) \qquad (4.21)$$

While the phase-shift angle has the effect of altering the shape of the composite waveform, the phase sequence order of the harmonics is not affected.

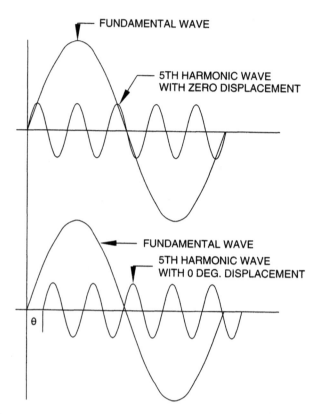

FUNDAMENTAL WAVE

5TH HARMONIC WAVE
WITH ZERO DISPLACEMENT

FUNDAMENTAL WAVE

5TH HARMONIC WAVE
WITH 0 DEG. DISPLACEMENT

θ

FIGURE 4.8 Nonsymmetry of the waveform with respect to a vertical reference plane introduced by a displacement of harmonics. Periodicity is still maintained.

4.5 CAUSES OF VOLTAGE AND CURRENT HARMONICS

A pure sinusoidal waveform with zero harmonic distortion is a hypothetical quantity
and not a practical one. The voltage waveform, even at the point of generation,
contains a small amount of distortion due to nonuniformity in the excitation magnetic
field and discrete spatial distribution of coils around the generator stator slots. The
distortion at the point of generation is usually very low, typically less than 1.0%.
The generated voltage is transmitted many hundreds of miles, transformed to several
levels, and ultimately distributed to the power user. The user equipment generates
currents that are rich in harmonic frequency components, especially in large commercial or industrial installations. As harmonic currents travel to the power source,
the current distortion results in additional voltage distortion due to impedance voltages associated with the various power distribution equipment, such as transmission
and distribution lines, transformers, cables, buses, and so on. Figure 4.9 illustrates
how current distortion is transformed into voltage distortion. Not all voltage distortion, however, is due to the flow of distorted current through the power system
impedance. For instance, static uninterruptible power source (UPS) systems can

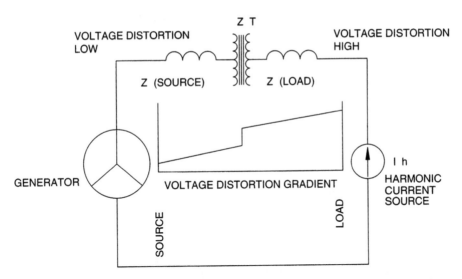

FIGURE 4.9 Voltage distortion due to current distortion. The gradient graph indicates how distortion changes from source to load.

generate appreciable voltage distortion due to the nature of their operation. Normal AC voltage is converted to DC and then reconverted to AC in the inverter section of the UPS. Unless waveform shaping circuitry is provided, the voltage waveforms generated in UPS units tend to be distorted.

As nonlinear loads are propagated into the power system, voltage distortions are introduced which become greater moving from the source to the load because of the circuit impedances. Current distortions for the most part are caused by loads. Even loads that are linear will generate nonlinear currents if the supply voltage waveform is significantly distorted. When several power users share a common power line, the voltage distortion produced by harmonic current injection of one user can affect the other users. This is why standards are being issued that will limit the amount of harmonic currents that individual power users can feed into the source (an issue that we will examine later in this chapter). The major causes of current distortion are nonlinear loads due to adjustable speed drives, fluorescent lighting, rectifier banks, computer and data-processing loads, arc furnaces, and so on. One can easily visualize an environment where a wide spectrum of harmonic frequencies are generated and transmitted to other loads or other power users, thereby producing undesirable results throughout the system.

4.6 INDIVIDUAL AND TOTAL HARMONIC DISTORTION

Individual harmonic distortion (IHD) is the ratio between the root mean square (RMS) value of the individual harmonic and the RMS value of the fundamental

$$IHD_n = I_n/I_1 \qquad\qquad (4.22)$$

For example, assume that the RMS value of the third harmonic current in a nonlinear load is 20 A, the RMS value of the fifth harmonic current is 15 A, and the RMS value of the fundamental is 60 A. Then, the individual third harmonic distortion is:

$$IHD_3 = 20/60 = 0.333, \text{ or } 33.3\%$$

and the individual fifth harmonic distortion is:

$$IHD_5 = 15/60 = 0.25, \text{ or } 25.0\%$$

Under this definition, the value of IHD_1 is always 100%. This method of quantifying the harmonics is known as harmonic distortion based on the fundamental. This is the convention used by the Institute of Electrical and Electronic Engineers (IEEE) in the U.S. The European International Electrotechnical Commission (IEC) quantifies harmonics based on the total RMS value of the waveform. Using the same example shown above, the RMS value of the waveform is:

$$I_{rms} = \sqrt{(60^2 + 20^2 + 15^2)} = 65 \text{ A}$$

Based on the IEC convention,

$$IHD_1 = 60/65 = 0.923, \text{ or } 92.3\%$$

$$IHD_3 = 20/65 = 0.308, \text{ or } 30.8\%$$

$$IHD_5 = 15/65 = 0.231, \text{ or } 23.1\%$$

The examples illustrate that even though the magnitudes of the harmonic currents are the same, the distortion percentages are different because of a change in the definition. It should be pointed out that it really does not matter what convention is used as long as the same one is maintained throughout the harmonic analysis. In this book, the IEEE convention will be followed, and all harmonic distortion calculations will be based on the fundamental.

Total harmonic distortion (THD) is a term used to describe the net deviation of a nonlinear waveform from ideal sine waveform characteristics. Total harmonic distortion is the ratio between the RMS value of the harmonics and the RMS value of the fundamental. For example, if a nonlinear current has a fundamental component of I_1 and harmonic components of $I_2, I_3, I_4, I_5, I_6, I_7, \ldots$, then the RMS value of the harmonics is:

$$I_H = \sqrt{(I_2^2 + I_3^2 + I_4^2 + I_5^2 + I_6^2 + I_7^2 + \ldots)} \qquad (4.23)$$

$$THD = (I_H/I_1) \times 100\% \qquad (4.24)$$

Example: Find the total harmonic distortion of a voltage waveform with the following harmonic frequency make up:

$$\text{Fundamental} = V_1 = 114 \text{ V}$$

$$\text{3rd harmonic} = V_3 = 4 \text{ V}$$

$$\text{5th harmonic} = V_5 = 2 \text{ V}$$

$$\text{7th harmonic} = V_7 = 1.5 \text{ V}$$

$$\text{9th harmonic} = V_9 = 1 \text{ V}$$

This problem can be solved in two ways:

$$\text{RMS value of the harmonics} = V_H = \sqrt{(4^2 + 2^2 + 1.5^2 + 1^2)} = 4.82 \text{ V}$$

$$THD = (4.82/114) \times 100 \cong 4.23\%$$

or find the individual harmonic distortions:

$$IHD_3 = 4/114 = 3.51\%$$

$$IHD_5 = 2/114 = 1.75\%$$

$$IHD_7 = 1.5/114 = 1.32\%$$

$$IHD_9 = 1/114 = 0.88\%$$

By definition, $IHD_1 = 100\%$, so

$$THD = \sqrt{(IHD_3^2 + IHD_5^2 + IHD_7^2 + IHD_9^2)} \cong 4.23\%$$

The results are not altered by using either the magnitude of the RMS quantities or the individual harmonic distortion values.

The individual harmonic distortion indicates the contribution of each harmonic frequency to the distorted waveform, and the total harmonic distortion describes the net deviation due to all the harmonics. These are both important parameters. In order to solve harmonic problems, we require information on the composition of the individual distortions so that any treatment may be tailored to suit the problem. The total harmonic distortion, while conveying no information on the harmonic makeup, is used to describe the degree of pollution of the power system as far as harmonics are concerned. Defining the individual and total harmonic distortions will be helpful as we look at some typical nonlinear waveforms and their harmonic frequency characteristics.

4.7 HARMONIC SIGNATURES

Many of the loads installed in present-day power systems are harmonic current generators. Combined with the impedance of the electrical system, the loads also produce harmonic voltages. The nonlinear loads may therefore be viewed as both harmonic current generators and harmonic voltage generators. Prior to the 1970s, speed control of AC motors was primarily achieved using belts and pulleys. Now, adjustable speed drives (ASDs) perform speed control functions very efficiently. ASDs are generators of large harmonic currents. Fluorescent lights use less electrical energy for the same light output as incandescent lighting but produce substantial harmonic currents in the process. The explosion of personal computer use has resulted in harmonic current proliferation in commercial buildings. This section is devoted to describing, in no particular order, a few of the more common nonlinear loads that surround us in our everyday life.

4.7.1 FLUORESCENT LIGHTING

Figure 4.10 shows a current waveform at a distribution panel supplying exclusively fluorescent lights. The waveform is primarily comprised of the third and the fifth harmonic frequencies. The individual current harmonic distortion makeup is provided in Table 4.2. The waveform also contains slight traces of even harmonics, especially of the higher frequency order. The current waveform is flat topped due to initiation of arc within the gas tube, which causes the voltage across the tube and the current to become essentially unchanged for a portion of each half of a cycle.

4.7.2 ADJUSTABLE SPEED DRIVES

While several technologies exist for creating a variable voltage and variable frequency power source for the speed control of AC motors, the pulse-width modulation (PWM) drive technology is currently the most widely used. Figures 4.11 and 4.12 show current graphs at the ASD input lines with a motor operating at 60 and 45 Hz, respectively. Tables 4.3 and 4.4 show the harmonic current distortion spectrum for the two respective frequencies. The characteristic double hump for each half cycle of the AC waveform is due to conduction of the input rectifier modules for a duration of two 60° periods for each half cycle. As the operating frequency is reduced, the humps become pronounced with a large increase in the total harmonic distortion. The THD of 74.2% for 45-Hz operation is excessive and can produce many deleterious effects, as will be shown in later sections of this chapter.

Figure 4.13 is the waveform of the voltage at the ASD input power lines. It was stated earlier that large current distortions can produce significant voltage distortions. In this particular case, the voltage THD is 8.3%, which is higher than levels typically found in most industrial installations. High levels of voltage THD also produce unwanted results. Table 4.5 provides the voltage harmonic distortion distribution.

Figure 4.14 is the current waveform of an ASD of smaller horsepower. This drive contains line side inductors which, along with the higher inductance of the motor, produce a current waveform with less distortion. Table 4.6 provides the harmonic frequency distribution for this ASD.

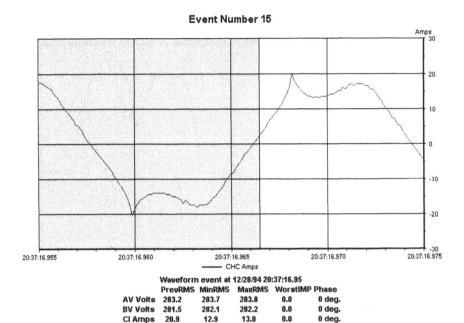

FIGURE 4.10 Nonlinear current drawn by fluorescent lighting.

TABLE 4.2
Harmonic Number *h(n)* vs. Individual Harmonic
Distortion (IHD) for a Fluorescent Lighting Load

Harmonic Distortion Spectrum					
h(n)	IHD (%)	*h(n)*	IHD (%)	*h(n)*	IHD (%)
0	—	11	2.2	22	0.6
1	100	12	0.3	23	0.6
2	0.3	13	1.7	24	0.7
3	13.9	14	0.3	25	1.4
4	0.3	15	1.9	26	1.1
5	9	16	0.3	27	0.3
6	0.2	17	0.8	28	0.9
7	3.3	18	0.5	29	1.5
8	0	19	1.4	30	1
9	3.2	20	0.4	31	0.3
10	0.1	21	1.2		

Note: Total harmonic distortion = 18.0%.

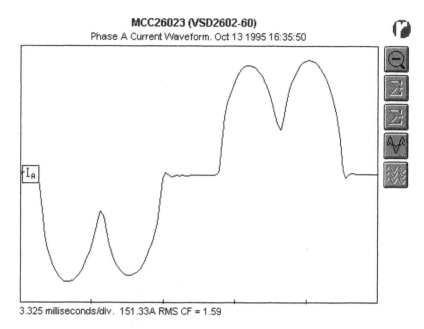

FIGURE 4.11 Adjustable speed drive input current with motor operating at 60 Hz.

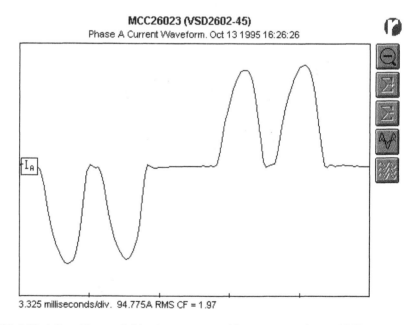

FIGURE 4.12 Adjustable speed drive input current with motor operating at 45 Hz.

TABLE 4.3
Harmonic Number $h(n)$ vs. Individual Harmonic
Distortion for an Adjustable Speed Drive Input Current
with Motor Running at 60 Hz

		Harmonic Distortion Spectrum			
$h(n)$	IHD (%)	$h(n)$	IHD (%)	$h(n)$	IHD (%)
0	0.15	11	9.99	22	0.39
1	100	12	0.03	23	2.95
2	4.12	13	0.19	24	0.02
3	0.78	14	0.48	25	0.66
4	1.79	15	0.07	26	0.15
5	35.01	16	0.52	27	0.05
6	0.215	17	4.85	28	0.22
7	2.62	18	0.03	29	1.79
8	1	19	0.67	30	0.03
9	0.06	20	0.31	31	0.64
10	0.73	21	0.04		

Note: Total harmonic distortion = 37.3%.

TABLE 4.4
Harmonic Number $h(n)$ vs. Individual Harmonic
Distortion for an Adjustable Speed Drive Input Current
with Motor Running at 45 Hz

		Harmonic Distortion Spectrum			
$h(n)$	IHD (%)	$h(n)$	IHD (%)	$h(n)$	IHD (%)
0	2.23	11	6.36	22	0.16
1	100	12	0.03	23	3.75
2	4.56	13	9.99	24	0.12
3	2.44	14	0.11	25	1.73
4	3.29	15	0.62	26	0.42
5	62.9	16	0.35	27	0.33
6	1.4	17	5.22	28	0.22
7	36.1	18	0.35	29	1.68
8	0.43	19	1.96	30	0.26
9	0.73	20	0.64	31	1.36
10	0.58	21	0.22		

Note: Total harmonic distortion = 74.2%.

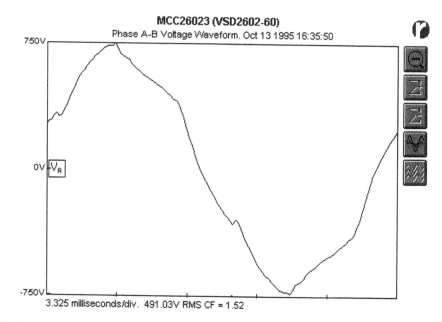

FIGURE 4.13 Adjustable speed drive input voltage with motor operating at 60 Hz.

TABLE 4.5
Harmonic Number $h(n)$ vs. Individual Harmonic Distortion for an Adjustable Speed Drive Input Voltage with Motor Running at 60 Hz

		Harmonic Distortion Spectrum			
$h(n)$	IHD (%)	$h(n)$	IHD (%)	$h(n)$	IHD (%)
0	0.02	11	1.87	22	0.07
1	100	12	0.02	23	0.46
2	0.12	13	0.92	24	0.04
3	0.09	14	0.07	25	0.36
4	0.11	15	0.01	26	0.06
5	7.82	16	0.04	27	0.03
6	0.01	17	0.61	28	0.07
7	1.42	18	0.06	29	0.4
8	0.06	19	0.36	30	0.02
9	0.04	20	0.06	31	0.34
10	0.03	21	0.12		

Note: Total harmonic distortion = 8.3%.

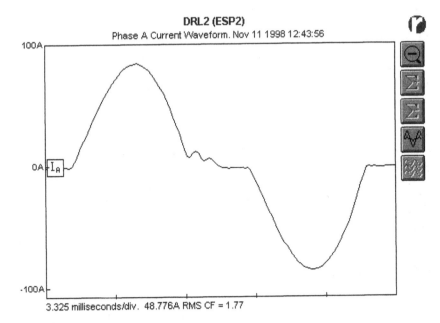

FIGURE 4.14 Adjustable speed drive input current for a smaller motor operating at 50 Hz (ASD with input line inductors).

TABLE 4.6
Harmonic Number *h*(*n*) vs. Individual Harmonic Distortion for an Adjustable Speed Drive Input Voltage with Line Inductor and Motor Running at 50 Hz

Harmonic Distortion Spectrum					
h(*n*)	IHD (%)	*h*(*n*)	IHD (%)	*h*(*n*)	IHD (%)
0	1.27	11	0.93	22	0.43
1	100	12	0.44	23	0.42
2	1.76	13	1.01	24	0.29
3	35.5	14	0.35	25	0.51
4	1.91	15	0.96	26	0.24
5	3.83	16	0.53	27	0.58
6	1.62	17	0.23	28	0.15
7	3.42	18	0.64	29	0.2
8	0.93	19	0.82	30	0.13
9	3.22	20	0.44	31	0.21
10	0.54	21	0.75		

Note: Total harmonic distortion = 36.3%.

4.7.3 PERSONAL COMPUTER AND MONITOR

Figures 4.15 and 4.16 show the nonlinear current characteristics of a personal computer and a computer monitor, respectively. Tables 4.7 and 4.8 provide the harmonic content of the currents for the two devices. The predominance of the third and fifth harmonics is evident. The current THD for both devices exceeds 100%, as the result of high levels of individual distortions introduced by the third and fifth harmonics. The total current drawn by a personal computer and its monitor is less than 2 A, but a typical high-rise building can contain several hundred computers and monitors. The net effect of this on the total current harmonic distortion of a facility is not difficult to visualize.

So far we have examined some of the more common harmonic current generators. The examples illustrate that a wide spectrum of harmonic currents is generated. Depending on the size of the power source and the harmonic current makeup, the composite harmonic picture will be different from facility to facility.

4.8 EFFECT OF HARMONICS ON POWER SYSTEM DEVICES

We are interested in the subject of harmonics because of the harmful effects they have on power system devices. What makes harmonics so insidious is that very often

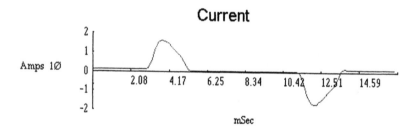

FIGURE 4.15 Nonlinear current drawn by single personal computer.

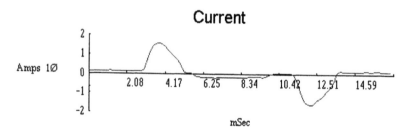

FIGURE 4.16 Nonlinear current drawn by single computer video monitor.

the effects of harmonics are not known until failure occurs. Insight into how harmonics can interact within a power system and how they can affect power system components is important for preventing failures. In this section, we will look at the effect of harmonics on some common power system devices.

TABLE 4.7
Harmonic Number h(n) vs. Individual Harmonic Distortion for a Personal Computer

Harmonic Distortion Spectrum					
h(n)	IHD (%)	h(n)	IHD (%)	h(n)	IHD (%)
0	12.8	11	10.3	22	2.1
1	100	12	1.2	23	0
2	3.3	13	10.3	24	0
3	87.2	14	0	25	0
4	5.1	15	10.3	26	0
5	64.1	16	0	27	0
6	1.6	17	5.1	28	0
7	41.1	18	0	29	0
8	0	19	2.4	30	0
9	17.9	20	0	31	0
10	1.1	21	2.1		

Note: Total harmonic distortion = 118.3%.

TABLE 4.8
Harmonic Frequency h(n) vs. Individual Harmonic Distortion for Computer Monitor Current

Harmonic Distortion Spectrum					
h(n)	IHD (%)	h(n)	IHD (%)	h(n)	IHD (%)
0	0	11	10	22	0
1	100	12	2.5	23	5
2	5	13	10	24	0
3	90	14	2.5	25	0
4	5	15	10	26	0
5	62.5	16	0	27	0
6	5	17	2.5	28	0
7	32.5	18	0	29	0
8	0	19	0	30	0
9	12.5	20	0	31	0
10	2.5	21	5		

Note: Total harmonic distortion = 116.3%.

4.8.1 TRANSFORMERS

Harmonics can affect transformers primarily in two ways. Voltage harmonics produce additional losses in the transformer core as the higher frequency harmonic voltages set up hysteresis loops, which superimpose on the fundamental loop. Each loop represents higher magnetization power requirements and higher core losses. A second and a more serious effect of harmonics is due to harmonic frequency currents in the transformer windings. The harmonic currents increase the net RMS current flowing in the transformer windings which results in additional I^2R losses. Winding eddy current losses are also increased. Winding eddy currents are circulating currents induced in the conductors by the leakage magnetic flux. Eddy current concentrations are higher at the ends of the windings due to the crowding effect of the leakage magnetic field at the coil extremities. The winding eddy current losses increase as the square of the harmonic current and the square of the frequency of the current. Thus, the eddy loss (EC) is proportional to $I_h^2 \times h^2$, where I_h is the RMS value of the harmonic current of order h, and h is the harmonic frequency order or number. Eddy currents due to harmonics can significantly increase the transformer winding temperature. Transformers that are required to supply large nonlinear loads must be derated to handle the harmonics. This derating factor is based on the percentage of the harmonic currents in the load and the rated winding eddy current losses.

One method by which transformers may be rated for suitability to handle harmonic loads is by k factor ratings. The k factor is equal to the sum of the square of the harmonic frequency currents (expressed as a ratio of the total RMS current) multiplied by the square of the harmonic frequency numbers:

$$k = I_1^2(1)^2 + I_2^2(2)^2 + I_3^2(3)^2 + I_4^2(4)^2 + \dots + I_n^2(n)^2 \qquad (4.25)$$

where

I_1 is the ratio between the fundamental current and the total RMS current.
I_2 is the ratio between the second harmonic current and the total RMS current.
I_3 is the ratio between the third harmonic current and the total RMS current.

Equation (4.25) can be rewritten as:

$$k = \Sigma \, I_n^2 h^2 (h = 1, 2, 3, \dots, n) \qquad (4.26)$$

Example: Determine the k rating of a transformer required to carry a load consisting of 500 A of fundamental, 200 A of third harmonics, 120 A of fifth harmonics, and 90 A of seventh harmonics:

$$\text{Total RMS current } (I) = \sqrt{(500^2 + 200^2 + 120^2 + 90^2)} = 559 \text{ A}$$

$$I_1 = 500/559 = 0.894$$

$$I_3 = 200/559 = 0.358$$

$$I_5 = 120/559 = 0.215$$

$$I_7 = 90/559 = 0.161$$

$$k = (0.894)^2 1^2 + (0.358)^2 3^2 + (0.215)^2 5^2 + (0.161)^2 7^2 = 4.378$$

The transformer specified should be capable of handling 559 A of total RMS current with a k factor of not less than 4.378. Typically, transformers are marked with k ratings of 4, 9, 13, 20, 30, 40, and 50, so a transformer with a k rating of 9 should be chosen. Such a transformer would have the capability to carry the full RMS load current and handle winding eddy current losses equal to k times the normal rated eddy current losses.

The k factor concept is derived from the ANSI/IEEE C57.110 standard, Recommended Practices for Establishing Transformer Capability When Supplying Non-Sinusoidal Load Currents, which provides the following expression for derating a transformer when supplying harmonic loads:

$$I \text{ max.(pu)} = [P_{LL-R(pu)}/1 + (\Sigma f_h^2 h^2/\Sigma f_h^2)P_{EC-R(pu)}]^{1/2} \tag{4.27}$$

where

I max.(pu) = ratio of the maximum nonlinear current of a specified harmonic makeup that the transformer can handle to the transformer rated current.

$P_{LL-R(pu)}$ = load loss density under rated conditions (per unit of rated load I^2R loss density.

$P_{EC-R(pu)}$ = winding eddy current loss under rated conditions (per unit of rated I^2R loss).

f_h = harmonic current distribution factor for harmonic h (equal to harmonic h current divided by the fundamental frequency current for any given load level).

h = harmonic number or order.

As difficult as this formula might seem, the underlying principle is to account for the increased winding eddy current losses due to the harmonics. The following example might help clarify the IEEE expression for derating a transformer.

Example: A transformer with a full load current rating of 1000 A is subjected to a load with the following nonlinear characteristics. The transformer has a rated winding eddy current loss density of 10.0% (0.10 pu). Find the transformer derating factor.

Harmonic number (h)	f_h (pu)
1	1
3	0.35
5	0.17
7	0.09

Maximum load loss density, $P_{LL-R(pu)} = 1 + 0.1 = 1.1$

Maximum rated eddy current loss density, $P_{EC-R(pu)} = 0.1$

$\Sigma f_h^2 h^2 = 1^2 + (0.35)^2 3^2 + (0.17)^2 5^2 + (0.09)^2 7^2 = 3.22$

$\Sigma f_h^2 = 1^2 + 0.35^2 + 0.17^2 + 0.09^2 = 1.16$

I max.(pu) $= [1.1/1 + (3.22 \times 0.1/1.16)]^{1/2} = 0.928$

The transformer derating factor is 0.928; that is, the maximum nonlinear current of the specified harmonic makeup that the transformer can handle is 928 A.

The ANSI/IEEE derating method is very useful when it is necessary to calculate the allowable maximum currents when the harmonic makeup of the load is known. For example, the load harmonic conditions might change on an existing transformer depending on the characteristics of new or replacement equipment. In such cases, the transformer may require derating. Also, transformers that supply large third harmonic generating loads should have the neutrals oversized. This is because, as we saw earlier, the third harmonic currents of the three phases are in phase and therefore tend to add in the neutral circuit. In theory, the neutral current can be as high as 173% of the phase currents. Transformers for such applications should have a neutral bus that is twice as large as the phase bus.

4.8.2 AC MOTORS

Application of distorted voltage to a motor results in additional losses in the magnetic core of the motor. Hysteresis and eddy current losses in the core increase as higher frequency harmonic voltages are impressed on the motor windings. Hysteresis losses increase with frequency and eddy current losses increase as the square of the frequency. Also, harmonic currents produce additional $I^2 R$ losses in the motor windings which must be accounted for.

Another effect, and perhaps a more serious one, is torsional oscillations due to harmonics. Table 4.1 classified harmonics into one of three categories. Two of the more prominent harmonics found in a typical power system are the fifth and seventh harmonics. The fifth harmonic is a negative sequence harmonic, and the resulting magnetic field revolves in a direction opposite to that of the fundamental field at a speed five times the fundamental. The seventh harmonic is a positive sequence harmonic with a resulting magnetic field revolving in the same direction as the fundamental field at a speed seven times the fundamental. The net effect is a magnetic field that revolves at a relative speed of six times the speed of the rotor. This induces currents in the rotor bars at a frequency of six times the fundamental frequency. The resulting interaction between the magnetic fields and the rotor-induced currents produces torsional oscillations of the motor shaft. If the frequency of the oscillation coincides with the natural frequency of the motor rotating members, severe damage to the motor can result. Excessive vibration and noise in a motor operating in a harmonic environment should be investigated to prevent failures.

Motors intended for operation in a severe harmonic environment must be specially designed for the application. Motor manufacturers provide motors for operation with ASD units. If the harmonic levels become excessive, filters may be applied at the motor terminals to keep the harmonic currents from the motor windings. Large motors supplied from ASDs are usually provided with harmonic filters to prevent motor damage due to harmonics.

4.8.3 CAPACITOR BANKS

Capacitor banks are commonly found in commercial and industrial power systems to correct for low power factor conditions. Capacitor banks are designed to operate at a maximum voltage of 110% of their rated voltages and at 135% of their rated kVARS. When large levels of voltage and current harmonics are present, the ratings are quite often exceeded, resulting in failures. Because the reactance of a capacitor bank is inversely proportional to frequency, harmonic currents can find their way into a capacitor bank. The capacitor bank acts as a sink, absorbing stray harmonic currents and causing overloads and subsequent failure of the bank.

A more serious condition with potential for substantial damage occurs due to a phenomenon called harmonic resonance. Resonance conditions are created when the inductive and capacitive reactances become equal at one of the harmonic frequencies. The two types of resonances are series and parallel. In general, series resonance produces voltage amplification and parallel resonance results in current multiplication. Resonance will not be analyzed in this book, but many textbooks on electrical circuit theory are available that can be consulted for further explanation. In a harmonic-rich environment, both series and parallel resonance may be present. If a high level of harmonic voltage or current corresponding to the resonance frequency exists in a power system, considerable damage to the capacitor bank as well as other power system devices can result. The following example might help to illustrate power system resonance due to capacitor banks.

Example: Figure 4.17 shows a 2000-kVA, 13.8-kV to 480/277-V transformer with a leakage reactance of 6.0% feeding a bus containing two 500-hp adjustable speed drives. A 750-kVAR Y-connected capacitor bank is installed on the 480-V bus for power factor correction. Perform an analysis to determine the conditions for resonance (consult Figure 4.18 for the transformer and capacitor connections and their respective voltages and currents):

Transformer secondary current $(I) = 2000 \times 10^3 / \sqrt{3 \times 480} = 2406$ A

Transformer secondary volts $= (V) = 277$

Transformer reactance $= I \times X_L \times 100/V = 6.0$

Transformer leakage reactance $(X_L) = 0.06 \times 277/2406 = 0.0069$ Ω

$X_L = 2\pi f L$, where $L = 0.0069/377 = 0.183 \times 10^{-4}$ H

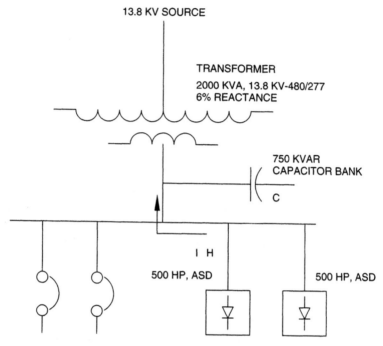

FIGURE 4.17 Schematic representation of an adjustable speed drive and a capacitor bank supplied from a 2000-kVA power transformer.

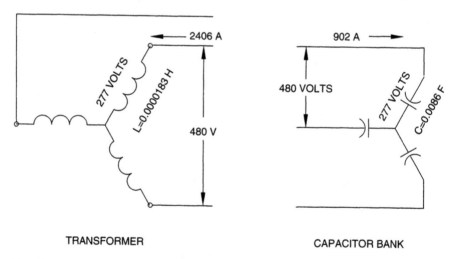

FIGURE 4.18 Transformer and capacitor bank configuration.

For the capacitor bank,

$$\sqrt{3} \times 480 \times I_C = 750 \times 10^3, \text{ where } I_C = 902 \text{ A}$$

$$\text{Capacitive reactance } (X_C) = V/I_C = 277/902 = 0.307 \text{ }\Omega$$

$$X_C = 1/2\pi f C, \text{ where } C = 1/(377 \times 0.307) = 86 \times 10^{-4} \text{ F}$$

For resonance, $X_L = X_C$; therefore,

$$2\pi f_R L = 1/2\pi f_R C$$

where f_R is the resonance frequency

$$f_R = 1/2\pi \sqrt{LC} \cong 401 \text{ Hz}$$

The resonance frequency is 401 Hz or the 6.7th (401/60) harmonic frequency. The resonance frequency is close to the seventh harmonic frequency, which is one of the more common harmonic frequency components found in power systems. This condition can have very serious effects.

The following expression presents a different way to find the harmonic resonance frequency:

$$\text{Resonance frequency order} = R_n = \sqrt{(MVA_{SC} \div MVAR_C)} \qquad (4.28)$$

where MVA_{SC} is the available symmetrical fault MVA at the point of connection of the capacitor in the power system, and $MVAR_C$ is the rating of the capacitor bank in $MVAR$. In the above example, neglecting the source impedance, the available fault current = $2406 \div 0.06 \cong 40,100$ A.

$$\text{Available fault } MVA = \sqrt{3} \times 480 \times 40,100 \times 10^{-6} = 33.34$$

$$\text{Capacitor } MVAR = 0.75$$

Therefore, the resonance frequency number = $\sqrt{33.34 \div 0.75} = 6.67$, and the harmonic frequency = $6.67 \times 60 = 400.2$. This proves that similar results are obtained by using Eq. (4.28). The expression in Eq. (4.28) is derived as follows: The available three-phase fault current at the common bus is given by $I_{SC} = V \div X$, where V is the phase voltage in kilovolts and X is the total reactance of the power system at the bus. I_{SC} is in units of kiloamperes.

$$I_{SC} = V \div 2\pi f_1 L, \text{ where } f_1 \text{ is the fundamental frequency}$$

$$\text{Short circuit } MVA = MVA_{SC} = 3 \times V \times I_{SC} = 3V^2 \div 2\pi f_1 L$$

From this,

$$L = 3V^2 \div 2\pi f_1 (MVA_{SC})$$

At resonance,

$$X_{LR} = 2\pi f_R L = 3V^2 f_R \div f_1 (MVA_{SC})$$

Because $f_R \div f_1$ = resonance frequency order, R_n, then

$$X_{LR} = 3V^2 R_n \div (MVA_{SC})$$

For the capacitor bank, $I_C = V \div X_C$, and capacitor reactive power $MVAR_C = 3 \times V \times I_C = 3V^2(2\pi f_1 C)$. We can derive an expression for the capacitive reactance at resonance $X_{CR} = 3V^2 \div R_n(MVAR_C)$. Equating X_{LR} and X_{CR}, the harmonic order at resonance is the expression given by Eq. (4.28).

The capacitor bank and the transformer form a parallel resonant circuit with the seventh harmonic current from the ASDs acting as the harmonic source. This condition is represented in Figure 4.19. Two adjustable speed drives typically draw a current of 550 A each, for a total load of 1100 A. If the seventh harmonic current is 5.0% of the fundamental (which is typical in drive applications), the seventh harmonic current seen by the parallel resonant circuit is 55 A = I_7.

If the resistance of the transformer and the associated cable, bus, etc. is 1.0%, then $R \cong 0.0012 \ \Omega$.

The quality factor, Q, of an electrical system is a measure of the energy stored in the inductance and the capacitance of the system. The current amplification factor (CAF) of a parallel resonance circuit is approximately equal to the Q of the circuit:

$$Q = 2\pi (\text{maximum energy stored}) / \text{energy dissipated per cycle}$$

$$Q = (2\pi)(1/2)LI_m^2 \div (I^2R)/f, \text{ where } I_m = \sqrt{2}I$$

$$Q = X/R$$

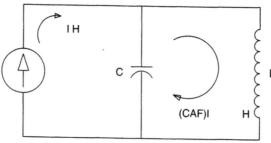

FIGURE 4.19 Parallel resonance circuit formed by transformer inductance and capacitor bank capacitance at harmonic frequency f_H.

For the seventh harmonic frequency, $CAF = X_7/R = 7 \times 0.0069/0.0012 = 40.25$. Therefore, current $I_R = 40.25 \times 55 = 2214$ A. The net current through the capacitor bank $= \sqrt{(I_C^2 + I_R^2)} = 2390$ A. It is easy to see that the capacitor bank is severely overloaded. If the capacitor protective device does not operate to isolate the bank, the capacitor bank will be damaged.

In the above example, by changing the capacitor bank to a 500-kVAR unit, the resonance frequency is increased to 490 Hz, or the 8.2 harmonic. This frequency is potentially less troublesome. (The reader is encouraged to work out the calculations.) In addition, the transformer and the capacitor bank may also form a series resonance circuit as viewed from the power source. This condition can cause a large voltage rise on the 480-V bus with unwanted results. Prior to installing a capacitor bank, it is important to perform a harmonic analysis to ensure that resonance frequencies do not coincide with any of the characteristic harmonic frequencies of the power system.

4.8.4 CABLES

Current flowing in a cable produces I^2R losses. When the load current contains harmonic content, additional losses are introduced. To compound the problem, the effective resistance of the cable increases with frequency because of the phenomenon known as skin effect. Skin effect is due to unequal flux linkage across the cross section of the conductor which causes AC currents to flow only on the outer periphery of the conductor. This has the effect of increasing the resistance of the conductor for AC currents. The higher the frequency of the current, the greater the tendency of the current to crowd at the outer periphery of the conductor and the greater the effective resistance for that frequency.

The capacity of a cable to carry nonlinear loads may be determined as follows. The skin effect factor is calculated first. The skin effect factor depends on the skin depth, which is an indicator of the penetration of the current in a conductor. Skin depth (δ) is inversely proportional to the square root of the frequency:

$$\delta = S \div \sqrt{f}$$

where S is a proportionality constant based on the physical characteristics of the cable and its magnetic permeability and f is the frequency of the current.

If R_{dc} is the DC resistance of the cable, then the AC resistance at frequency f, $(R_f) = K \times R_{dc}$. The value of K is determined from Table 4.9 according to the value of X, which is calculated as:

$$X = 0.0636 \sqrt{f\mu \div R_{dc}} \tag{4.29}$$

where 0.0636 is a constant for copper conductors, f is the frequency, μ is the magnetic permeability of the conductor material, and R_{dc} is the DC resistance per mile of the conductor. The magnetic permeability of a nonmagnetic material such as copper is approximately equal to 1.0. Tables or graphs containing values of X and K are available from cable manufacturers.

TABLE 4.9
Cable Skin Effect Factor

X	K	X	K	X	K
0	1	1.4	1.01969	2.7	1.22753
0.1	1	1.5	1.02558	2.8	1.2662
0.2	1	1.6	1.03323	2.9	1.28644
0.3	1.00004	1.7	1.04205	3.0	1.31809
0.5	1.00032	1.8	1.0524	3.1	1.35102
0.6	1.00067	1.9	1.0644	3.1	1.38504
0.7	1.00124	2.0	1.07816	3.3	1.41999
0.8	1.00212	2.1	1.09375	3.4	1.4577
0.9	1.0034	2.1	1.11126	3.5	1.49202
1.0	1.00519	2.3	1.13069	3.6	1.52879
1.1	1.00758	2.4	1.15207	3.7	1.56587
1.2	1.01071	2.5	1.17538	3.8	1.60312
1.3	1.0147	2.6	1.20056	3.9	1.64051

Example: Find the 60-Hz and 420-Hz resistance of a 4/0 copper cable with a DC resistance of 0.276 Ω per mile. Using Eq. (4.29),

$$X_{60} = 0.0636 \sqrt{(60 \times 1 \div 0.276)} = 0.938$$

From Table 4.2, $K \cong 1.004$, and $R_{60} = 1.004 \times 0.276 = 0.277$ Ω per mile. Also,

$$X_{420} = 0.0636 \sqrt{(420 \times 1 \div 0.276)} = 2.48$$

From Table 4.2, $K \cong 1.154$, and $R_{420} = 1.154 \times 0.276 = 0.319$ Ω per mile.

The ratio of the resistance of the cable at a given frequency to its resistance at 60 Hz is defined as the skin effect ratio, E. According to this definition,

$E_2 =$ resistance at second harmonic frequency ÷ resistance at the fundamental frequency $= R_{120} \div R_{60}$

$E_3 =$ resistance at third harmonic frequency ÷ resistance at the fundamental frequency $= R_{180} \div R_{60}$

Also, remember that the general form expression for the individual harmonic distortions states that I_n is equal to the RMS value of the nth harmonic current divided by the RMS value of the fundamental current, thus an expression for the current rating factor for cables can be formulated. The current rating factor (q) is the equivalent fundamental frequency current at which the cable should be rated for carrying nonlinear loads containing harmonic frequency components:

$$q = I_1^2 E_1 + I_2^2 E_2 + I_3^2 E_3 + \ldots + I_n^2 E_n \qquad (4.30)$$

where I_1, I_2, I_3 are the ratios of the harmonic frequency currents to the fundamental current, and E_1, E_2, E_3 are the skin effect ratios.

Example: Determine the current rating factor for a 300-kcmil copper conductor required to carry a nonlinear load with the following harmonic frequency content:

$$\text{Fundamental} = 250 \text{ A}$$

$$\text{3rd harmonic} = 25 \text{ A}$$

$$\text{5th harmonic} = 60 \text{ A}$$

$$\text{7th harmonic} = 45 \text{ A}$$

$$\text{11th harmonic} = 20 \text{ A}$$

The DC resistance of 300-kcmil cable = 0.17 Ω per mile. Using Eq. (4.29),

$$X_{60} = 0.0636 \sqrt{(60 \times 1 \div 0.17)} = 1.195, K \cong 1.0106$$

$$X_{180} = 0.0636 \sqrt{(180 \times 1 \div 0.17)} = 2.069, K \cong 1.089$$

$$X_{300} = 0.0636 \sqrt{(300 \times 1 \div 0.17)} = 2.672, K \cong 1.220$$

$$X_{420} = 0.0636 \sqrt{(420 \times 1 \div 0.17)} = 3.161, K \cong 1.372$$

$$X_{660} = 0,0636 \sqrt{(660 \times 1 \div 0.17)} = 3.963, K \cong 1.664$$

$$R_{60} = 1.0106 \times 0.17 = 0.1718 \ \Omega/\text{mile}$$

$$R_{180} = 1.089 \times 0.17 = 0.1851 \ \Omega/\text{mile}$$

$$R_{300} = 1.220 \times 0.17 = 0.2074 \ \Omega/\text{mile}$$

$$R_{420} = 1.372 \times 0.17 = 0.2332 \ \Omega/\text{mile}$$

$$R_{660} = 1.664 \times 0.17 = 0.2829 \ \Omega/\text{mile}$$

Skin effect ratios are:

$$E_1 = 1, E_3 = 1.077, E_5 = 1.207, E_7 = 1.357, E_{11} = 1.647$$

The individual harmonic distortion factors are:

$$I_1 = 1.0, I_3 = 25/250 = 0.1, I_5 = 60/250 = 0.24, I_7 = 0.18, I_{11} = 20/250 = 0.08$$

The current rating factor from Eq. (4.30) is given by:

$$q = 1 + (0.1)^2(1.077) + (0.24)^2(1.207) + (0.18)^2(1.357) + (0.08)^2(1.647) = 1.135$$

The cable should be capable of handling a 60-Hz equivalent current of $1.135 \times 250 \cong 284$ A.

4.8.5 BUSWAYS

Most commercial multistory installations contain busways that serve as the primary source of electrical power to various floors. Busways that incorporate sandwiched busbars are susceptible to nonlinear loading, especially if the neutral bus carries large levels of triplen harmonic currents (third, ninth, etc.). Under the worst possible conditions, the neutral bus may be forced to carry a current equal to 173% of the phase currents. In cases where substantial neutral currents are expected, the busways must be suitably derated. Table 4.10 indicates the amount of nonlinear loads that may be allowed to flow in the phase busbars for different neutral currents. The data are shown for busways with neutral busbars that are 100 and 200% in size.

TABLE 4.10
Bus Duct Derating Factor
for Harmonic Loading

$I_N/I_{\varnothing H}$	$I_{\varnothing H}/I_\varnothing$	
	100% N	200% N
0	1.000	1.000
0.25	0.99	0.995
0.50	0.961	0.98
0.75	0.918	0.956
1.00	0.866	0.926
1.25	0.811	0.891
1.50	0.756	0.853
1.75	0.703	0.814
2.00	0.655	0.775

Note: I_N is the neutral current, $I_{\varnothing H}$ is the harmonic current component in each phase, and $I_\varnothing$ is the total phase current. N = size of neutral bus bar in relation to phase bus bar.

4.8.6 PROTECTIVE DEVICES

Harmonic currents influence the operation of protective devices. Fuses and motor thermal overload devices are prone to nuisance operation when subjected to nonlinear currents. This factor should be given due consideration when sizing protective devices for use in a harmonic environment. Electromechanical relays are also affected by harmonics. Depending on the design, an electromechanical relay may operate faster or slower than the expected times for operation at the fundamental frequency alone. Such factors should be carefully considered prior to placing the relays in service.

4.9 GUIDELINES FOR HARMONIC VOLTAGE AND CURRENT LIMITATION

So far we have discussed the adverse effects of harmonics on power system operation. It is important, therefore, that attempts be made to limit the harmonic distortion that a facility might produce. There are two reasons for this. First, the lower the harmonic currents produced in an electrical system, the better the equipment within the confinement of the system will perform. Also, lower harmonic currents produce less of an impact on other power users sharing the same power lines of the harmonic generating power system. The IEEE 519 standard provides guidelines for harmonic current limits at the point of common coupling (PCC) between the facility and the utility. The rationale behind the use of the PCC as the reference location is simple. It is a given fact that within a particular power use environment, harmonic currents will be generated and propagated. Harmonic current injection at the PCC determines how one facility might affect other power users and the utility that supplies the power. Table 4.11 (per IEEE 519) lists harmonic current limits based on the size of the power user. As the ratio between the maximum available short circuit current at the PCC and the maximum demand load current increases, the percentage of the harmonic currents that are allowed also increases. This means that larger power users are allowed to inject into the system only a minimal amount of harmonic current (as a percentage of the fundamental current). Such a scheme tends to equalize the amounts of harmonic currents that large and small users of power are allowed to inject into the power system at the PCC.

IEEE 519 also provides guidelines for maximum voltage distortion at the PCC (see Table 4.12). Limiting the voltage distortion at the PCC is the concern of the utility. It can be expected that as long as a facility's harmonic current contribution is within the IEEE 519 limits the voltage distortion at the PCC will also be within the specified limits.

TABLE 4.11
Harmonic Current Limits for General Distribution Systems (120–69,000 V)

I_{SC}/I_L	$h < 11$	$11 \leq h < 17$	$17 \leq h < 23$	$23 \leq h < 35$	$35 \leq h$	THD
<20	4.0	2.0	1.5	0.6	0.3	5.0
20–50	7.0	3.5	2.5	1.0	0.5	8.0
50–100	10.0	4.5	4.0	1.5	0.7	12.0
100–1000	12.0	5.5	5.0	2.0	1.0	15.0
>1000	15.0	7.0	6.0	2.5	1.4	20.0

Note: I_{SC} = maximum short-circuit current at PCC; I_L = maximum fundamental frequency demand load current at PCC (average current of the maximum demand for the preceding 12 months); h = individual harmonic order; THD = total harmonic distortion. based on the maximum demand load current. The table applies to odd harmonics; even harmonics are limited to 25% of the odd harmonic limits shown above.

TABLE 4.12
Voltage Harmonic Distortion Limits

Bus Voltage at PCC	Individual Voltage Distortion (%)	Total Voltage Distortion THD (%)
69 kV and below	3.0	5.0
69.001 kV through 161 kV	1.5	2.5
161.001 kV and above	1.0	1.5

Note: PCC = point of common coupling; THD = total harmonic distortion.

When the IEEE 519 harmonic limits are used as guidelines within a facility, the PCC is the common junction between the harmonic generating loads and other electrical equipment in the power system. It is expected that applying IEEE guidelines renders power system operation more reliable. In the future, more and more utilities might require facilities to limit their harmonic current injection to levels stipulated by IEEE 519. The following section contains information on how harmonic mitigation might be achieved.

4.10 HARMONIC CURRENT MITIGATION

4.10.1 EQUIPMENT DESIGN

The use of electronic power devices is steadily increasing. It is estimated that more than 70% of the loading of a facility by year 2010 will be due to nonlinear loads, thus demand is increasing for product manufacturers to produce devices that generate lower distortion. The importance of equipment design in minimizing harmonic current production has taken on greater importance, as reflected by technological improvements in fluorescent lamp ballasts, adjustable speed drives, battery chargers, and uninterruptible power source (UPS) units. Computers and similar data-processing devices contain switching mode power supplies that generate a substantial amount of harmonic currents, as seen earlier. Designing power supplies for electronic equipment adds considerably to the cost of the units and can also make the equipment heavier. At this time, when computer prices are extremely competitive, attempts to engineer power supplies that draw low harmonic currents are not a priority.

Adjustable speed drive (ASD) technology is evolving steadily, with greater emphasis being placed on a reduction in harmonic currents. Older generation ASDs using current source inverter (CSI) and voltage source inverter (VSI) technologies produced considerable harmonic frequency currents. The significant harmonic frequency currents generated in power conversion equipment can be stated as:

$$n = kq \pm 1$$

where n is the significant harmonic frequency, k is any positive integer (1, 2, 3, etc.), and q is the pulse number of the power conversion equipment which is the number of power pulses that are in one complete sequence of power conversion. For example, a three-phase full wave bridge rectifier has six power pulses and therefore has a pulse number of 6. With six-pulse-power conversion equipment, the following significant harmonics may be generated:

For $k = 1$, $n = (1 \times 6) \pm 1 = $ 5th and 7th harmonics.
For $k = 2$, $n = (2 \times 6) \pm 1 = $ 11th and 13th harmonics.

With six-pulse-power conversion equipment, harmonics below the 5th harmonic are insignificant. Also, as the harmonic number increases, the individual harmonic distortions become lower due to increasing impedance presented to higher frequency components by the power system inductive reactance. So, typically, for six-pulse-power conversion equipment, the 5th harmonic current would be the highest, the 7th would be lower than the 5th, the 11th would be lower than the 7th, and so on, as shown below:

$$I_{13} < I_{11} < I_7 < I_5$$

We can deduce that, when using 12-pulse-power conversion equipment, harmonics below the 11th harmonic can be made insignificant. The total harmonic distortion is also considerably reduced. Twelve-pulse-power conversion equipment costs more than six-pulse-power equipment. Where harmonic currents are the primary concern, 24-pulse-power conversion equipment may be considered.

4.10.2 HARMONIC CURRENT CANCELLATION

Transformer connections employing phase shift are sometimes used to effect cancellation of harmonic currents in a power system. Triplen harmonic (3rd, 9th, 15th, etc.) currents are a set of currents that can be effectively trapped using a special transformer configuration called the zigzag connection. In power systems, triplen harmonics add in the neutral circuit, as these currents are in phase. Using a zigzag connection, the triplens can be effectively kept away from the source. Figure 4.20 illustrates how this is accomplished.

The transformer phase-shifting principle is also used to achieve cancellation of the 5th and the 7th harmonic currents. Using a Δ–Δ and a Δ–Y transformer to supply harmonic producing loads in parallel as shown in Figure 4.21, the 5th and the 7th harmonics are canceled at the point of common connection. This is due to the 30° phase shift between the two transformer connections. As the result of this, the source does not see any significant amount of the 5th and 7th harmonics. If the nonlinear loads supplied by the two transformers are identical, then maximum harmonic current cancellation takes place; otherwise, some 5th and 7th harmonic currents would still be present. Other phase-shifting methods may be used to cancel higher harmonics if they are found to be a problem. Some transformer manufacturers offer multiple phase-shifting connections in a single package which saves cost and space compared to using individual transformers.

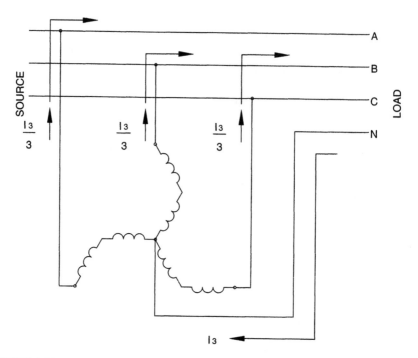

FIGURE 4.20 Zig-zag transformer application as third harmonic filter.

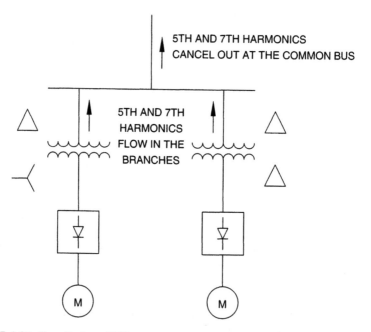

FIGURE 4.21 Cancellation of fifth and seventh harmonic currents by using 30° phase-shifted transformer connections.

4.10.3 HARMONIC FILTERS

Nonlinear loads produce harmonic currents that can travel to other locations in the power system and eventually back to the source. As we saw earlier, harmonic currents can produce a variety of effects that are harmful to the power system. Harmonic currents are a result of the characteristics of particular loads. As long as we choose to employ those loads, we must deal with the reality that harmonic currents will exist to a degree dependent upon the loads. One means of ensuring that harmonic currents produced by a nonlinear current source will not unduly interfere with the rest of the power system is to filter out the harmonics. Application of harmonic filters helps to accomplish this.

Harmonic filters are broadly classified into passive and active filters. Passive filters, as the name implies, use passive components such as resistors, inductors, and capacitors. A combination of passive components is tuned to the harmonic frequency that is to be filtered. Figure 4.22 is a typical series-tuned filter. Here the values of the inductor and the capacitor are chosen to present a low impedance to the harmonic frequency that is to be filtered out. Due to the lower impedance of the filter in comparison to the impedance of the source, the harmonic frequency current will circulate between the load and the filter. This keeps the harmonic current of the desired frequency away from the source and other loads in the power system. If other harmonic frequencies are to be filtered out, additional tuned filters are applied in parallel. Applications such as arc furnaces require multiple harmonic filters, as they generate large quantities of harmonic currents at several frequencies.

Applying harmonic filters requires careful consideration. Series-tuned filters appear to be of low impedance to harmonic currents but they also form a parallel resonance circuit with the source impedance. In some instances, a situation can be created that is worse than the condition being corrected. It is imperative that computer simulations of the entire power system be performed prior to applying harmonic filters. As a first step in the computer simulation, the power system is modeled to indicate the locations of the harmonic sources, then hypothetical harmonic filters are placed in the model and the response of the power system to the filter is examined. If unacceptable results are obtained, the location and values of the filter parameters are changed until the results are satisfactory. When applying harmonic filters, the units are almost never tuned to the exact harmonic frequency. For example, the 5th harmonic frequency may be designed for resonance at the 4.7th harmonic frequency.

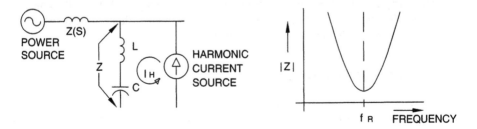

FIGURE 4.22 Series-tuned filter and filter frequency response.

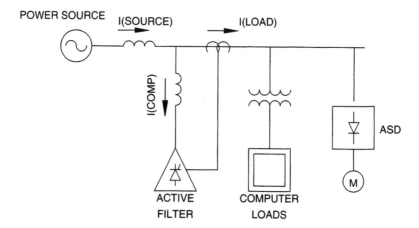

FIGURE 4.23 Active filter to cancel harmonic currents.

By not creating a resonance circuit at precisely the 5th harmonic frequency, we can minimize the possibility of the filter resonating with other loads or the source, thus forming a parallel resonance circuit at the 5th harmonic. The 4.7th harmonic filter would still be effective in filtering out the 5th harmonic currents. This is evident from the series-tuned frequency vs. impedance curve shown in Figure 4.22.

Sometimes, tuned filters are configured to provide power factor correction for a facility as well as harmonic current filtering. In such cases the filter would be designed to carry the resonant harmonic frequency current and also the normal frequency current at the fundamental frequency. In either case, a power system harmonic study is paramount to ensure that no ill effects would be produced by the application of the power factor correction/filter circuit.

Active filters use active conditioning to compensate for harmonic currents in a power system. Figure 4.23 shows an active filter applied in a harmonic environment. The filter samples the distorted current and, using power electronic switching devices, draws a current from the source of such magnitude, frequency composition, and phase shift to cancel the harmonics in the load. The result is that the current drawn from the source is free of harmonics. An advantage of active filters over passive filters is that the active filters can respond to changing load and harmonic conditions, whereas passive filters are fixed in their harmonic response. As we saw earlier, application of passive filters requires careful analysis. Active filters have no serious ill effects associated with them. However, active filters are expensive and not suited for application in small facilities.

4.11 CONCLUSIONS

The term *harmonics* is becoming very common in power systems, small, medium, or large. As the use of power electronic devices grows, so will the need to understand the effects of harmonics and the application of mitigation methods. Fortunately, harmonics in a strict sense are not transient phenomena. Their presence can be easily measured and identified. In some cases, harmonics can be lived with indefinitely, but in other cases they should be minimized or eliminated. Either of these approaches requires a clear understanding of the theory behind harmonics.

5   Grounding and Bonding

5.1 INTRODUCTION

What do the terms *grounding* and *bonding* mean? Quite often the terms are mistakenly used interchangeably. To reinforce understanding of the two concepts, the definitions given in Chapter 1 are repeated here: *Grounding* is a conducting connection by which an electrical circuit or equipment is connected to earth or to some conducting body of relatively large extent that serves in place of earth. *Bonding* is intentional electrical interconnecting of conductive paths in order to ensure common electrical potential between the bonded parts.

The primary purpose of grounding and bonding is electrical safety, but does *safety* cover personal protection or equipment protection, or both? Most people would equate electrical safety with personal protection (and rightfully so), but equipment protection may be viewed as an extension of personal protection. An electrical device grounded so that it totally eliminates shock hazards but could still conceivably start a fire is not a total personal protective system. This is why even though personal safety is the prime concern, equipment protection is also worthy of consideration when configuring a grounding system methodology.

With the advent of the electronic age, grounding and bonding have taken on the additional roles of serving as reference planes for low-level analog or digital signals. Two microelectronic devices that communicate with each other and interpret data require a common reference point from which to operate. The ground plane for such devices should provide a low-impedance reference plane for the devices, and any electrical noise induced or propagated to the ground plane should have very minimal impact on the devices. So far we have identified three reasons for grounding and bonding. One point that cannot be stressed enough is that nothing that we do to grounding and bonding should compromise personal safety. It is not uncommon to see modifications to the grounding of an electrical system for the sole purpose of making equipment function properly at the expense of safety. Such actions contradict the real reason for grounding a system in the first place.

5.2 SHOCK AND FIRE HAZARDS

Grounding and bonding of electrical devices and systems are vital to ensuring that people living or working in the environment will be adequately protected. We will start by looking into why personal safety is dependent on grounding and bonding. Table 5.1 is a list of physiological hazards associated with passage of electrical current through an average human body. It is obvious that it does not take much current to cause injury and even death. The resistance of an average human under conditions when the skin is dry is about 100 kΩ or higher. When the skin is wet,

TABLE 5.1
Effect of Current Flow Through Human Body

Current Level	Shock Hazard
100 μA	Threshold of perception
1–5 mA	Sensation of pain
5–10 mA	Increased pain
10–20 mA	Intense pain; unable to release grip
30 mA	Breathing affected
40–60 mA	Feeling of asphyxiation
75 mA	Ventricular fibrilation, irregular heartbeat

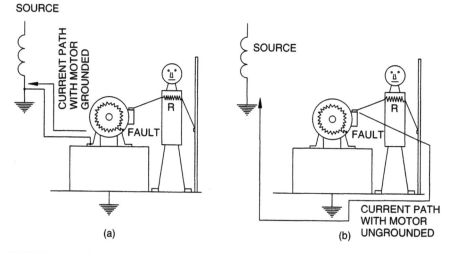

FIGURE 5.1 (a) Current flow when motor frame is grounded. (b) Current flow when motor frame is ungrounded.

the resistance drops to 10 kΩ or lower. It is not difficult to see how susceptible humans are to shock hazard.

Figures 5.1a and 5.1b illustrate what would happen if a person came in contact with the frame of an electric motor where, due to insulation deterioration, a 480-V phase is in contact with the frame. Figure 5.1a is the condition of the frame being bonded to the service ground terminal, which in turn is connected to the building ground electrode. If the power source feeding the motor is a grounded source, this condition in all likelihood would cause the overcurrent protection (such as the fuse or circuit breaker) to operate and open the circuit to the motor. If the power source feeding the motor is an ungrounded source (such as a Δ-connected transformer), no overcurrent protection is likely to operate; however, the phase that is contacting the frame will be brought to the ground potential and the person touching the frame is not in danger of receiving an electric shock.

On the other hand, consider Figure 5.1b, where the motor frame is not bonded to a ground. If the source feeding the motor were a grounded source, considerable leakage current would flow through the body of the person. The current levels can reach values high enough to cause death. If the source is ungrounded, the current flow through the body will be completed by the stray capacitance of cable used to connect the motor to the source. For a 1/0 cable the stray capacitance is of the order of 0.17 μF for a 100-ft cable. The cable reactance is approximately 15,600 Ω. Currents significant enough to cause a shock would flow through the person in contact with the motor body.

5.3 NATIONAL ELECTRICAL CODE GROUNDING REQUIREMENTS

Grounding of electrical systems is mandated by the electrical codes that govern the operation of electrical power systems. The National Electrical Code (NEC) in the U.S. is the body that lays out requirements for electrical systems for premises. However, the NEC does not cover installations in ships, railways, or aircraft or underground in mines or electrical installations under the exclusive control of utilities.

Article 250 of the NEC requires that the following electrical systems of 50 to 1000 V should be grounded:

- Systems that can be grounded so that the maximum voltage to ground does not exceed 150 V
- Three-phase, four-wire, Wye-connected systems in which the neutral is used as a circuit conductor
- Three-phase, four-wire, Δ-connected systems in which the midpoint of one phase winding is used as a circuit conductor

Alternating current systems of 50 to 1000 V that should be permitted to be grounded but are not required to be grounded by the NEC include:

- Electrical systems used exclusively to supply industrial electric furnaces for melting, refining, tempering, and the like
- Separately derived systems used exclusively for rectifiers that supply adjustable speed industrial drives
- Separately derived systems supplied by transformers that have a primary voltage rating less than 1000 V, provided all of the following conditions are met:
  - The system is used exclusively for industrial controls.
  - The conditions of maintenance and supervision ensure that only qualified personnel will service the installation.
  - Continuity of control power is required.
  - Ground detectors are installed in the control system.

Article 250 of the NEC also states requirements for grounding for systems less than 50 V and those rated 1000 V and higher; interested readers are urged to refer to the Article.

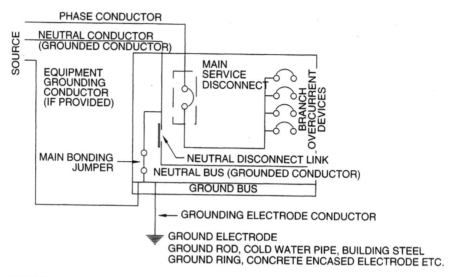

FIGURE 5.2 Main service switchboard indicating elements of a ground system.

5.4 ESSENTIALS OF A GROUNDED SYSTEM

Figure 5.2 shows the essential elements of a grounded electrical power system. It is best to have a clear understanding of the components of a ground system to fully grasp the importance of grounding for safety and power quality. The elements of Figure 5.2 are defined as follows:

Grounded conductor: A circuit conductor that is intentionally grounded (for example, the neutral of a three-phase Wye connected system or the midpoint of a single-phase 240/120 V system)

Grounding conductor: A conductor used to connect the grounded circuit of a system to a grounding electrode or electrodes

Equipment grounding conductor: Conductor used to connect the non-current-carrying metal parts of equipment, raceways, and other enclosures to the system grounded conductor, the grounding electrode conductor, or both at the service equipment or at the source of a separately derived system

Grounding electrode conductor: Conductor used to connect the grounding electrode to the equipment grounding conductor, the grounded conductor, or both

Main bonding jumper: An unspliced connection used to connect the equipment grounding conductor and the service disconnect enclosure to the grounded conductor of a power system

Ground: Earth or some conducting body of relatively large extent that serves in place of the earth

Ground electrode: A conductor or body of conductors in intimate contact with the earth for the purpose of providing a connection with the ground

5.5 GROUND ELECTRODES

In this section, various types of ground electrodes and their use will be discussed. The NEC states that the following elements are part of a ground electrode system in a facility:

- Metal underground water pipe
- Metal frame of buildings or structures
- Concrete-encased electrodes
- Ground ring
- Other made electrodes, such as underground structures, rod and pipe electrodes, and plate electrodes, when none of the above-listed items is available.

The code prohibits the use of a metal underground gas piping system as a ground electrode. Also, aluminum electrodes are not permitted. The NEC also mentions that, when applicable, each of the items listed above should be bonded together. The purpose of this requirement is to ensure that the ground electrode system is large enough to present low impedance to the flow of fault energy. It should be recognized that, while any one of the ground electrodes may be adequate by itself, bonding all of these together provides a superior ground grid system.

Why all this preoccupation with ground systems that are extensive and inter-connected? The answer is low impedance reference. A facility may have several individual buildings, each with its own power source. Each building may even have several power sources, such as transformers, uninterruptible power source (UPS) units, and battery systems. It is important that the electrical system or systems of each building become part of the same overall grounding system. A low impedance ground reference plane results from this arrangement (Figure 5.3). Among the additional benefits to the creation of a low-impedance earth-ground system is the fact that when an overhead power line contacts the earth, a low-impedance system will help produce ground-fault currents of sufficient magnitude to operate the over-current protection. When electrical charges associated with lightning strike a building and its electrical system, the lightning energy could pass safely to earth without damaging electrical equipment or causing injury to people. It is the author's personal experience that a lack of attention to grounding and bonding has been responsible for many preventable accidents involving equipment and personnel.

5.6 EARTH RESISTANCE TESTS

The earth resistance test is a means to ensure that the ground electrode system of a facility has adequate contact with earth. Figure 5.4 shows how an earth resistance tester is used to test the resistance between the ground grid and earth. The most common method of testing earth resistance is the fall of potential test, for which the earth resistance tester is connected as shown in Figure 5.4. The ground electrode of the facility or building is used as the reference point. Two ground rods are driven as indicated. The farthest rod is called the current rod (C_2), and the rod at the

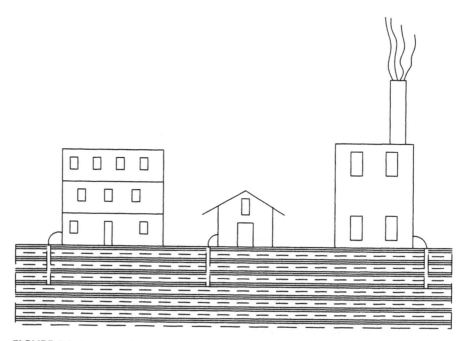

FIGURE 5.3 Low-impedance ground reference, provided by the earth, between several buildings in the same facility.

intermediate point is the potential rod (P_2). A known current is circulated between the reference electrode and the current rod. The voltage drop is measured between the reference ground electrode and the potential rod. The ground resistance is calculated as the ratio between the voltage and the current. The tester automatically calculates and displays the resistance in ohms. The potential rod is then moved to another location and the test repeated. The resistance values are plotted against the distance from the reference rod. The graph in Figure 5.4 is a typical earth resistance curve. The earth resistance is represented by the value corresponding to the flat portion of the curve. In typical ground grid systems, the value at a distance 62% of the total distance between the reference electrode and the current rod is taken as the resistance of the ground system with respect to earth.

The distance between the reference electrode and the current rod is determined by the type and size of the ground grid system. For a single ground rod, a distance of 100 to 150 ft is adequate. For large ground grid systems consisting of multiple ground rods, ground rings, or concrete-encased systems, the distance between the reference ground electrode and the current rod should be 5 to 10 times the diagonal measure of the ground grid system. The reason is that, as currents are injected into the earth, electrical fields are set up around the electrodes in the form of shells. To prevent erroneous results, the two sets of shells around the reference electrode and the current electrode should not overlap. The greater the distance between the two, the more accurate the ground resistance test results.

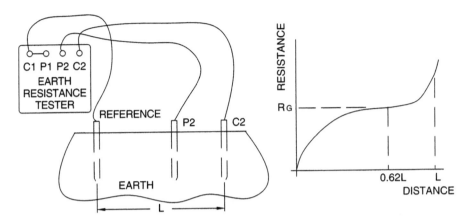

FIGURE 5.4 Ground resistance test instrument and plot of ground resistance and distance.

Article 250, Section 250-56, of the NEC code states that a single ground electrode that does not have a resistance of 25 Ω or less must be augmented by an additional electrode. Earth resistance of 25 Ω is adequate for residential and small commercial buildings. For large buildings and facilities that house sensitive loads, a resistance value of 10 Ω is typically specified. For buildings that contain sensitive loads such as signal, communication, and data-processing equipment, a resistance of 5 Ω or less is sometimes specified.

Earth resistance depends on the type of soil, its mineral composition, moisture content, and temperature. Table 5.2 provides the resistivity of various types of soils; Table 5.3, the effect of moisture on soil resistivity; and Table 5.4, the effect of temperature on soil resistivity. The information contained in the tables is used to illustrate the effect of various natural factors on soil resistivity. Table 5.5 shows the changes in earth resistance by using multiple ground rods. Note that, to realize the full benefits of multiple rods, the rods should be spaced an adequate distance apart.

TABLE 5.2
Resistivities of Common Materials

Material	Resistivity Range (Ω-cm)
Surface soils	100–5000
Clay	200–10,000
Sand and gravel	5000–100,000
Limestone	500–400,000
Shales	500–10,000
Sandstone	2000–200,000
Granite	1,000,000
Tap water	1000–5000
Seawater	20–200

TABLE 5.3
Effect of Moisture on Soil Resistivity

Moisture Content (% by weight)	Resistivity (Ω-cm)	
	Top Soil	Sandy Loam
0	1000×10^6	1000×10^6
2.5	250,000	15,000
5	165,000	43,000
10	53,000	22,000
15	21,000	13,000
20	12,000	10,000
30	10,000	8000

TABLE 5.4
Effect of Temperature on Earth Resistivity[a]

Temperature		Resistivity (Ω-cm)
°C	°F	
20	68	7200
10	50	9900
0	32 (water)	13,800
0	32 (ice)	30,000
−5	23	79,000
−15	5	330,000

[a] For sandy loam, 15.2% moisture.

TABLE 5.5
Change in Earth Resistance with Multiple Ground Rods

Number of Ground Rods	Distance between Rods[a]		
	D = L (%)	D = 2L (%)	D = 4L (%)
1	100	—	—
2	60	52	50
3	42	37	35
4	35	29	27
5	28	25	23
10	16	14	12

[a] One ground rod of length L is used as reference.

5.7 EARTH–GROUND GRID SYSTEMS

Ground grids can take different forms and shapes. The ultimate purpose is to provide a metal grid of sufficient area of contact with the earth so as to derive low resistance between the ground electrode and the earth. Two of the main requirements of any ground grid are to ensure that it will be stable with time and that it will not form chemical reactions with other metal objects in the vicinity, such as buried water pipes, building reinforcment bars, etc., and cause corrosion either in the ground grid or the neighboring metal objects.

5.7.1 GROUND RODS

According to the NEC, ground rods should be not less than 8 ft long and should consist of the following:

- Electrodes of conduits or pipes that are no smaller than 3/4-inch trade size; when these conduits are made of steel, the outer surface should be galvanized or otherwise metal-coated for corrosion protection
- Electrodes of rods of iron or steel that are at least 5/8 inches in diameter; the electrodes should be installed so that at least an 8-ft length is in contact with soil

Typically, copper-clad steel rods are used for ground rods. Steel provides the strength needed to withstand the forces during driving of the rod into the soil, while the copper coating provides corrosion protection and also allows copper conductors to be attached to the ground rod. The values indicated above are the minimum values; depending on the installation and the type of soil encountered, larger and longer rods or pipes may be needed. Table 5.6 shows earth resistance variation with the length of the ground rod, and Table 5.7 shows earth resistance values for ground rods of various diameters. The values are shown for a soil with a typical ground resistivity of 10,000 Ω-cm.

TABLE 5.6
Effect of Ground Rod Length on Earth Resistance

Ground Rod Length (ft)	Earth Resistance (Ω)
5	40
8	25
10	21
12	18
15	17

Note: Soil resistivity = 10,000 Ω-cm.

TABLE 5.7
Effect of Ground Rod Diameter on Earth Resistance[a]

Rod Diameter (inches)	% Resistance
0.5	100
0.75	90
1.0	85
1.5	78
2.0	76

Note: Soil resistivity = 10,000 Ω-cm.

[a] Resistance of a 0.5-inch-diameter rod is used as reference.

5.7.2 PLATES

Rectangular or circular plates should present an area of at least 2 ft² to the soil. Electrodes of iron and steel shall be at least 1/4 inch in thickness; electrodes of nonferrous metal should have a minimum thickness of 0.06 inch. Plate electrodes are to be installed at a minimum distance of 2.5 ft below the surface of the earth. Table 5.8 gives the earth resistance values for circular plates buried 3 ft below the surface in soil with a resistivity of 10,000 Ω-cm.

5.7.3 GROUND RING

The ground ring encircling a building in direct contact with the earth should be installed at a depth of not less than 2.5 ft below the surface of the earth. The ground ring should consist of at least 20 ft of bare copper conductor sized not less than #2 AWG. Typically, ground rings are installed in trenches around the building, and wire tails are brought out for connection to the grounded service conductor at the service disconnect panel or switchboard. It is preferred that a continuous piece of wire be

TABLE 5.8
Resistance of Circular Plates Buried 3 Feet Below Surface

Plate Area (ft²)	Earth Resistance (Ω)
2	30
4	23
6	18
10	15
20	12

Note: Soil resistivity = 10,000 Ω-cm.

TABLE 5.9
Earth Resistance of Buried Conductors

Wire Size	Resistance (Ω) for Total Buried Wire Length				
(AWG)	20 ft	40 ft	60 ft	100 ft	200 ft
# 6	23	14	7	5	3
# 1/0	18	12	6	4	2

Note: Soil resistivity = 10,000 Ω-cm.

used. In large buildings, this might be impractical. If wires are spliced together, the connections should be made using exothermic welding or listed wire connectors. Table 5.9 provides the resistance of two conductors buried 3 ft below the surface for various conductor lengths. The values contained in the table are intended to point out the variations that may be obtained using different types of earth electrodes. The values are not to be used for designing ground grids, as the values are apt to change with the type of soil and soil temperatures at the installation.

5.8 POWER GROUND SYSTEM

A good ground electrode grid system with low resistance to earth is a vital foundation for the entire power system for the facility. As we mentioned earlier, the primary objective of power system grounding is personal safety, in addition to limiting damage to equipment. When a ground fault occurs, large ground return currents are set up which causes the overcurrent protection to open and isolate the load from the power source. In many cases, the phase overcurrent protection is depended upon to perform this function during a ground fault. Article 250-95 of the NEC (1999) requires ground fault protection for solidly grounded Wye-connected electrical services of more than 150 V to ground, not exceeding 600 V phase-to-phase, for each service rated 1000 A or more. This requirement recognizes the need for ground fault protection for systems rated greater than 150 V to ground because of the possibility of arcing ground faults in such systems. Arcing ground faults generate considerably lower fault currents than bolted ground faults or direct short circuits between phase and ground. The possibility of arcing ground faults in systems rated less than 150 V to ground should be acknowledged, and ground fault protection against low-level ground faults should be provided for the power system. The ground fault protection is set at levels considerably lower than the phase fault protection. For instance, a 1000-A-rated overcurrent protection system may have the ground fault protection set at 200 A or lower. The setting of the ground fault device depends on the degree of protection required, as this requirement is strictly ground fault protection for equipment.

As indicated in Table 5.1, it takes very little current to cause electrical shock and even loss of life. This is why ground fault circuit interrupters (GFCIs) are required by the NEC for convenience outlets in certain areas of homes or facilities.

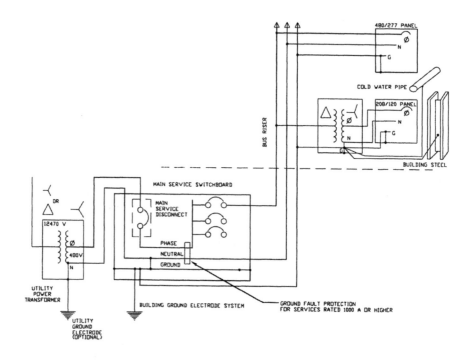

FIGURE 5.5 Typical power system grounding scheme.

GFCI protection is set to open a circuit at a current of 5 mA. The GFCI is not intended for equipment protection but is strictly for personal protection. Figure 5.5 illustrates a typical facility power-grounding scheme.

5.9 SIGNAL REFERENCE GROUND

Signal reference ground (SRG) is a relatively new term. The main purpose of the signal reference ground is not personal safety or equipment protection but merely to provide a common reference low-impedance plane from which sensitive loads may operate. Why is SRG important? Figure 5.6 depicts two low-level microcircuits sharing data and power lines. What makes this communication possible is that both devices have a common reference signal, the ground. If the reference ground is a high-impedance connection, voltage differentials may be created that would affect the point of reference for the two devices, so lowering the impedance between the reference points of the two circuits lowers the potential for coupling of noise between the devices.

When we mention low impedance, we mean low impedance at high frequencies. For power frequency, even a few hundred feet of wire can provide adequate imped-ance, but the situation is different at high frequencies. For example, let us consider

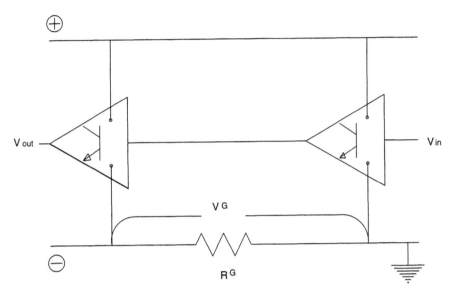

FIGURE 5.6 Ground potential difference due to excessive ground impedance.

a situation when two devices are connected to a 10-ft length of #4 copper conductor ground wire:

The DC resistance of the wire is = 0.00025 Ω
The inductive reactance at 60 Hz = 0.0012 Ω
Inductive reactance at 1 MHz $\cong$ 20 Ω

If a noise current of 100 mA at 1 MHz is to find its way into the ground wire between the two devices, the noise voltage must be 2 V, which is enough to cause the devices to lose communication and perhaps even sustain damage, depending on the device sensitivity. This example is a simple situation consisting of only two devices; however, hundreds and perhaps thousands of such devices or circuits might be present in an actual computer or communication data center. All these devices require a common reference from which to operate. This is accomplished by the use of the SRG.

As noted above, the main purpose of the SRG is not electrical safety, even though nothing that we do to the ground system should compromise safety; rather, the SRG is a ground plane that provides all sensitive equipment connected to it a reference point from which to operate without being unduly affected by noise that may be propagated through the SRG by devices external or internal to the space protected by the SRG. What we mean by this is that noise may be present in the SRG, but the presence of the noise should not result in voltage differentials or current loops of levels that could interfere with the operation of devices that use the SRG for reference.

The SRG is not a stand-alone entity; it must be bonded to other building ground electrodes such as building steel, ground ring, or concrete-encased electrodes. This requirement permits any noise impinging on the SRG to be safely conducted away from the SRG to building steel and the rest of the ground grid system.

5.10 SIGNAL REFERENCE GROUND METHODS

The SRG can take many forms, depending on the user preference. Some facilities use a single conductor installed underneath the floor and looped around the space of the computer center. While this method is practical, it is limited in functionality due to the large impedances associated with long wires, as mentioned earlier. Larger computer data centers use more than one conductor but the limitations are the same as stated above. A preferred SRG consists of #2 AWG or larger copper conductor laid underneath the floor of the computer or communication center to form a grid of 2 × 2-ft squares (Figure 5.7). By creating multiple parallel paths, the impedance for the reference plane is made equal for all devices and circuits sharing the SRG. If the impedance is measured at any two nodes of the SRG and plotted against frequency, the shape of the frequency characteristics would appear as shown in Figure 5.8. The impedance vs. frequency graph should appear the same across any two sets of nodes of the SRG, as this is the main objective of the SRG.

Some installations use copper strips instead of circular conductors to form the grid. Other facilities might use sheets of copper under the floor of the computer center as the SRG. Constructing an SRG with a continuous sheet of copper creates a reference plane made up of infinite parallel paths instead of a discrete number of parallel paths as with SRGs made up of circular wires. The SRG is also bonded to the building steel and the stanchions that support the raised floor of the computer center. Such an arrangement provides excellent noise immunity and allows the creation of a good reference plane for the sensitive circuits. Figure 5.9 depicts how an SRG for a large-sized computer center might be configured. Some installations use aboveground wiring methods instead of a raised-floor configuration. The principle behind the configuration of the SRG does not change whether the ground

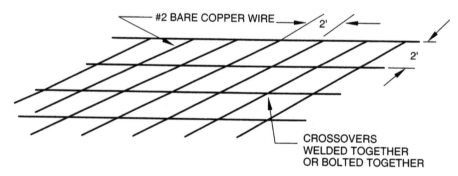

FIGURE 5.7 Typical 2 × 2-ft signal reference ground arrangement.

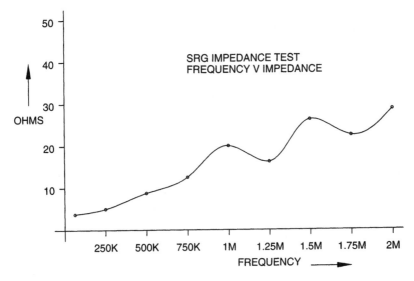

FIGURE 5.8 Typical signal reference ground frequency vs. impedance characteristics.

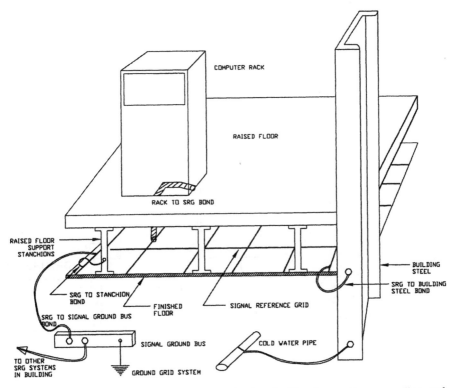

FIGURE 5.9 Typical computer and communication facility data center grounding and bonding.

reference plane is below ground or above ground. It is important that all noise-producing loads be kept away from the SRG. If such loads are present, they should be located at the outer periphery of the data center and bonded to the building steel, if possible.

5.11 SINGLE-POINT AND MULTIPOINT GROUNDING

With multipoint grounding, every piece of equipment sharing a common space or building is individually grounded (Figure 5.10); whereas, with single-point grounding, each piece of equipment is connected to a common bus or reference plane, which in turn is bonded to the building ground grid electrode (Figure 5.11). Multipoint grounding is adequate at power frequencies. For typical power systems, various transformers, UPS systems, and emergency generators located in each area or floor of the building are grounded to the nearest building ground electrode, such as building steel or coldwater pipe. Generally, this method is both convenient and economical, but it is neither effective nor recommended for grounding sensitive devices and circuits. As we saw, the primary purpose of grounding for sensitive equipment is the creation of a reference plane. This is best accomplished by single-point grounding and bonding means. The SRG must also be bonded to the building ground electrode to ensure personal safety.

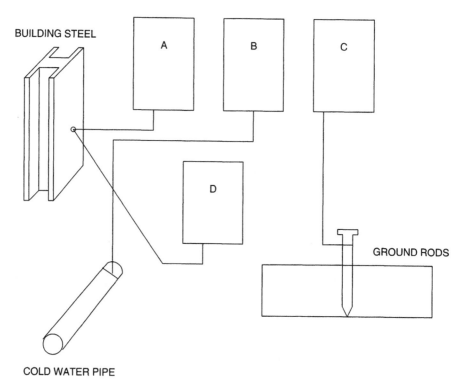

FIGURE 5.10 Multipoint ground system.

5.12 GROUND LOOPS

In Chapter 1, a ground loop was defined as a potentially detrimental loop formed when two or more points in an electrical system that are normally at ground potential are connected by a conducting path such that either or both points are not at the same potential. Let's examine the circuit shown in Figure 5.12. Here, the ground plane is at different potentials for the two devices that share the ground circuit. This sets up circulation of currents in the loop formed between the two devices by the common ground wires and the signal ground conductor. Such an occurrence can result in performance degradation or damage to devices within the loop. Ground loops are the result of faulty or improper facility wiring practices that cause stray currents to flow in the ground path, creating a voltage differential between two points in the ground system. They may also be due to a high-resistance or high-impedance connection between a device and the ground plane. Because the signal common or ground conductor is a low-impedance connection, it only takes a low-level ground loop potential to cause significant current to flow in the loop. By adhering to sound ground and bonding practices, as discussed throughout this chapter, ground loop potentials can be minimized or eliminated.

Problems due to ground loops can be difficult to identify and fix. The author has observed many instances where well-trained personnel have attempted to fix

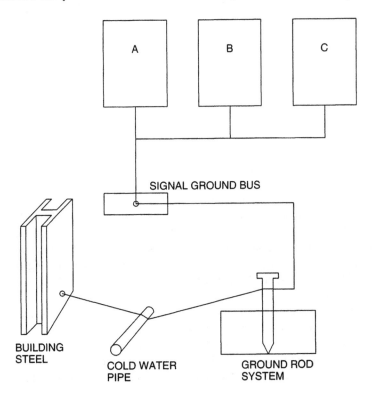

FIGURE 5.11 Single-point grounding of sensitive equipment.

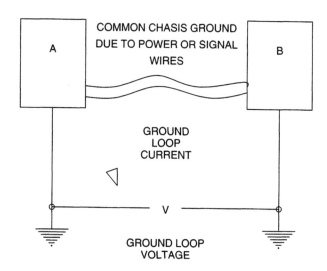

FIGURE 5.12 Ground loop voltage and current.

ground loop problems by removing the ground connections or ground pins from power and data cords. In all of these cases, relief, if any, has been minimal, and the conditions created by these actions are nothing short of lethal. This is what we mean by the statement that nothing that we do to grounding and bonding should make the installation unsafe.

5.13 ELECTROCHEMICAL REACTIONS DUE TO GROUND GRIDS

When two dissimilar metals are installed in damp or wet soil, an electrolytic cell is formed. If there is an external connection between the two metal members, a current can flow using the electrolyte formed by the wet soil which can cause deterioration of the anodic (+) member of the metal pair (Figure 5.13). The figure depicts the copper water pipes and ground rings bonded to the building steel or reinforcing bars in the foundation. This configuration results in current flow between the members. Over the course of time, the steel members that are more electropositive will start to disintegrate as they are asked to supply the electrons to support the current flow. If not detected, the structural integrity of the building is weakened. By suitably coating the steel or copper, current flow is interrupted and the electrolytic action is minimized.

Table 5.10 lists the metals in order of their position in the galvanic series. The more positive or anodic metals are more active and prone to corrosion. In some installations, to prevent corrosion of a specific metal member, sacrificial anodes are installed in the ground. The sacrificial anodes are more electropositive than the metals they are protecting, so they are sacrificed to protect the structural steel.

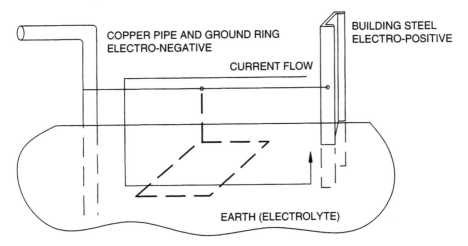

FIGURE 5.13 Metal corrosion due to electrolysis caused by copper and steel in the earth.

TABLE 5.10
Electromotive Series of Metals

Metal	Electrode Potential (V)
Magnesium	2.37
Aluminum	1.66
Zinc	0.763
Iron	0.44
Cadmium	0.403
Nickel	0.25
Tin	0.136
Lead	0.126
Copper	−0.337
Silver	−0.799
Palladium	−0.987
Gold	−1.5

5.14 EXAMPLES OF GROUNDING ANOMALIES OR PROBLEMS

5.14.1 LOSS OF GROUND CAUSES FATALITY

Case

At a manufacturing plant that used high-frequency, high-current welders for welding steel and aluminum parts, one of the welders took a break outdoors on a rainy day. When he walked back into the building and touched one of the welding machines to which power was turned on, he collapsed and died of cardiac arrest.

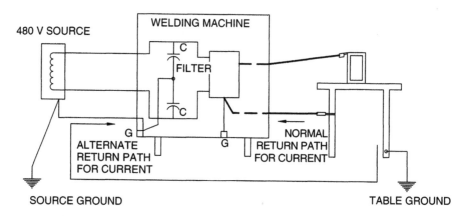

FIGURE 5.14 Example of grounding problem resulting in a fatality.

Investigation

Figure 5.14 shows the electrical arrangement of the equipment involved in this incident. The welding machine operated from a 480-V, one-phase source fed from a 480Y/277 secondary transformer. The input power lines of the machine contained a capacitance filter to filter high-frequency noise from the load side of the machine and to keep the noise from being propagated upstream toward the source. The return current for the welded piece was via the ground lead of the machine. Examination of the power and ground wiring throughout the building revealed several burned equipment ground wires. It was determined that due to improper or high-resistance connections in the return lead of the welder, the current was forced to return through the equipment ground wire of the machines. The equipment ground wires are not sized to handle large currents produced by the welding machine. This caused the equipment ground wire of the machine (and other machines) to be either severely damaged or totally burned off. As the center point of the capacitive filter is connected to the frame of the welding machine, the loss of ground caused the frame of the machine to float and be at a potential higher than the ground. When the operator touched the machine, he received a shock severe enough to cause cardiac arrest. The fact that he was exposed to moisture prior to the contact with the machine increased the severity of the hazard.

5.14.2 Stray Ground Loop Currents Cause Computer Damage

Case

In a commercial building, computers were burning up at an alarming rate. Most of the problems were found at the data ports.

Investigation

Wiring problems were found in the electrical panel supplying the computers, such as with the neutral wires in the ground terminal and ground wires in the neutral terminal. This configuration caused a portion of the neutral return current of the load

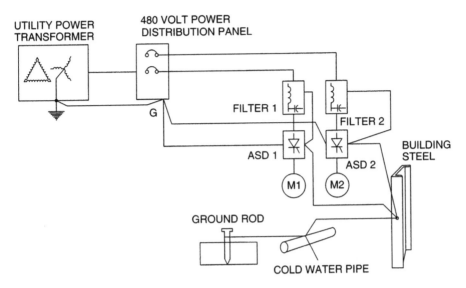

FIGURE 5.15 Adjustable speed drive grounding deficiencies, resulting in shutdowns and down time, reconfigured to correct the problem.

circuits to return via the equipment ground wires and other grounded parts such as conduits and water pipes. The current was also high in harmonic content, as would be expected in such an application. The flow of stray ground currents caused ground potential differences for the various computers that shared data lines. Resulting ground loop currents resulted in damage to data ports, which are not designed or intended to carry such currents. Once the wiring anomalies at the power distribution panel were corrected, computer damage was not experienced.

5.14.3 Ground Noise Causes Adjustable Speed Drives to Shut Down

Case

In a newspaper printing facility, two adjustable speed drives (ASDs) were installed as part of a new conveyor system to transport the finished product to the shipping area. The ASDs were shutting themselves off periodically, causing papers to back up on the conveyor and disrupting production.

Investigation

Figure 5.15 depicts the electrical setup of the ASDs. The electrical system in the facility was relatively old. The drives were supplied from a switchboard located some distance away. Tests revealed the presence of electrical noise in the lines supplying the ASDs. Even though the drives contained line filters, they were not effective in minimizing noise propagation. The conclusion was that the long length of the ground return wires for the drives presented high impedance to the noise, thereby allowing it to circulate in the power wiring. To correct the situation, the ASDs and the filters were bonded to building steel located close to the drives. The building steel was also bonded to the coldwater pipe and ground rods installed for this section of the power system. This created a good ground reference for the ASDs and the filter units. Noise was considerably minimized in the power wires. The drives operated satisfactorily after the changes were implemented.

5.15 CONCLUSIONS

A conclusion that we can draw is that grounding is not an area where one can afford to be lax. Reference is fundamental to the existence of stability. For electrical systems, reference is the ground or some other body large enough to serve in place of the ground, and electrical stability depends on how sound this reference is. We should not always think of this reference as a ground or a ground connected to the earth. For the electrical system of a ship, the hull of the ship and the water around the ship serve as the reference. For aircraft, the fuselage of the aircraft is the reference. Problems arise when we do not understand what the reference is for a particular application or we compromise the reference to try to make a system work. Either condition is a recipe for problems. Grounding is the foundation of any electrical power, communication, or data-processing system; when the foundation is taken care of, the rest of the system will be stable.

6   Power Factor

6.1 INTRODUCTION

Power factor is included in the discussion of power quality for several reasons. Power factor is a power quality issue in that low power factor can sometimes cause equipment to fail. In many instances, the cost of low power factor can be high; utilities penalize facilities that have low power factor because they find it difficult to meet the resulting demands for electrical energy. The study of power quality is about optimizing the performance of the power system at the lowest possible operating cost. Power factor is definitely an issue that qualifies on both counts.

6.2 ACTIVE AND REACTIVE POWER

Several different definitions and expressions can be applied to the term power factor, most of which are probably correct. Apparent power (S) in an electrical system can be defined as being equal to voltage times current:

$$S = V \times I(1\emptyset)$$

$$S = \sqrt{3} \times V \times I(3\emptyset)$$

where V = phase-to-phase voltage (V) and I = line current (VA).

Power factor (PF) may be viewed as the percentage of the total apparent power that is converted to real or useful power. Thus, active power (P) can be defined by:

$$P = V \times I \times PF - 1\emptyset$$

$$P = \sqrt{3} \times V \times I \times PF - 3\emptyset$$

In an electrical system, if the power factor is 0.80, 80% of the apparent power is converted into useful work. Apparent power is what the transformer that serves a home or business has to carry in order for that home or business to function. Active power is the portion of the apparent power that performs useful work and supplies losses in the electrical equipment that are associated with doing the work. Higher power factor leads to more optimum use of electrical current in a facility. Can a power factor reach 100%? In theory it can, but in practice it cannot without some form of power factor correction device. The reason why it can approach 100% power factor but not quite reach it is because all electrical circuits have inductance and capacitance, which introduce reactive power requirements. The reactive power is that

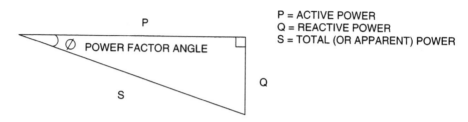

FIGURE 6.1 Power triangle and relationship among active, reactive, and apparent power.

portion of the apparent power that prevents it from obtaining a power factor of 100% and is the power that an AC electrical system requires in order to perform useful work in the system. Reactive power sets up a magnetic field in the motor so that a torque is produced. It is also the power that sets up a magnetic field in a transformer core allowing transfer of power from the primary to the secondary windings.

All reactive power requirements are not necessary in every situation. Any electrical circuit or device when subjected to an electrical potential develops a magnetic field that represents the inductance of the circuit or the device. As current flows in the circuit, the inductance produces a voltage that tends to oppose the current. This effect, known as Lenz's law, produces a voltage drop in the circuit that represents a loss in the circuit. At any rate, inductance in AC circuits is present whether it is needed or not. In an electrical circuit, the apparent and reactive powers are represented by the power triangle shown in Figure 6.1. The following relationships apply:

$$S = \sqrt{P^2 + Q^2} \tag{6.1}$$

$$P = S \cos\emptyset \tag{6.2}$$

$$Q = S \sin\emptyset \tag{6.3}$$

$$Q/P = \tan\emptyset \tag{6.4}$$

where S = apparent power, P = active power, Q = reactive power, and $\emptyset$ is the power factor angle. In Figure 6.2, V is the voltage applied to a circuit and I is the current in the circuit. In an inductive circuit, the current lags the voltage by angle $\emptyset$, as shown in the figure, and $\emptyset$ is called the power factor angle.

If X_L is the inductive reactance given by:

$$X_L = 2\pi f L$$

then total impedance (Z) is given by:

$$Z = R + jX_L$$

where j is the imaginary operator = $\sqrt{-1}$

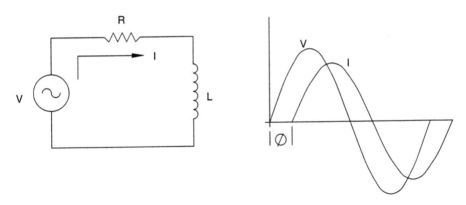

FIGURE 6.2 Voltage, current, and power factor angle in a resistive/inductive circuit.

The power factor angle is calculated from the expression:

$$\tan\emptyset = (X_L/R) \text{ or } \emptyset = \tan^{-1}(X_L/R) \tag{6.5}$$

Example: What is the power factor of a resistive/inductive circuit characterized by $R = 2\ \Omega$, $L = 2.0$ mH, $f = 60$ Hz?

$$X_L = 2\pi f L = 2 \times \pi \times 60 \times 2 \times 10^{-3} = 0.754\ \Omega$$

$$\tan\emptyset = X_L/R = 0.754/2 = 0.377$$

$$\emptyset = 20.66°$$

$$\text{Power factor} = PF = \cos(20.66) = 0.936$$

Example: What is the power factor of a resistance/capacitance circuit when $R = 10\ \Omega$, $C = 100\ \mu F$, and frequency $(f) = 60$ Hz? Here,

$$X_C = 1/2\pi f C = 1/2 \times \pi \times 60 \times 100 \times 10^{-6} = 26.54\ \Omega$$

$$\tan\emptyset = (-X_C/R) = -2.654$$

$$\emptyset = -69.35°$$

$$\text{Power factor} = PF = \cos\emptyset = 0.353$$

The negative power factor angle indicates that the current leads the voltage by 69.35°.

Let's now consider an inductive circuit where application of voltage V produces current I as shown in Figure 6.2 and the phasor diagram for a single-phase circuit is as shown. The current is divided into active and reactive components, I_P and I_Q:

$$I_p = I \times \cos\varnothing$$

$$I_Q = I \times \sin\varnothing$$

Active power $= P = V \times$ active current $= V \times I \times \cos\varnothing$

Reactive power $= Q = V \times$ reactive current $= V \times I \times \varnothing$

Total or apparent power $= S = \sqrt{(P^2 + Q^2)} = \sqrt{(V^2 I^2 \cos^2\varnothing + V^2 I^2 \sin^2\varnothing)} = V \times I$

Voltage, current, and power phasors are as shown in Figure 6.3. Depending on the reactive power component, the current phasor can swing, as shown in Figure 6.4. The ±90° current phasor displacement is the theoretical limit for purely inductive and capacitive loads with zero resistance, a condition that does not really exist in practice.

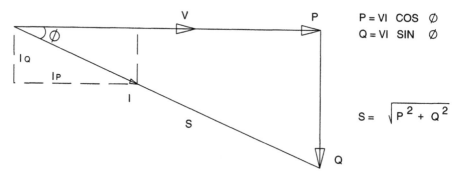

FIGURE 6.3 Relationship among voltage, current, and power phasors.

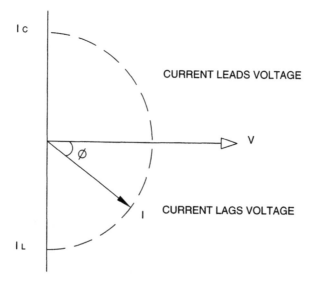

FIGURE 6.4 Theoretical limits of current.

6.3 DISPLACEMENT AND TRUE POWER FACTOR

The terms *displacement* and *true power factor,* are widely mentioned in power factor studies. Displacement power factor is the cosine of the angle between the fundamental voltage and current waveforms. The fundamental waveforms are by definition pure sinusoids. But, if the waveform distortion is due to harmonics (which is very often the case), the power factor angles are different than what would be for the fundamental waves alone. The presence of harmonics introduces additional phase shift between the voltage and the current. True power factor is calculated as the ratio between the total active power used in a circuit (including harmonics) and the total apparent power (including harmonics) supplied from the source:

True power factor = total active power/total apparent power

Utility penalties are based on the true power factor of a facility.

6.4 POWER FACTOR IMPROVEMENT

Two ways to improve the power factor and minimize the apparent power drawn from the power source are:

- Reduce the lagging reactive current demand of the loads
- Compensate for the lagging reactive current by supplying leading reactive current to the power system

The second method is the topic of interest in this chapter. Lagging reactive current represent the inductance of the power system and power system components. As observed earlier, lagging reactive current demand may not be totally eliminated but may be reduced by using power system devices or components designed to operate with low reactive current requirements. Practically no devices in a typical power system require leading reactive current to function; therefore, in order to produce leading currents certain devices must be inserted in a power system. These devices are referred to as *power factor correction equipment.*

6.5 POWER FACTOR CORRECTION

In simple terms, power factor correction means reduction of lagging reactive power (Q) or lagging reactive current (I_Q). Consider Figure 6.5. The source V supplies the resistive/inductive load with impedance (Z):

$$Z = R + j\omega L$$

$$I = V/Z = V/(R + j\omega L)$$

Apparent power $= S = V \times I = V^2/(R + j\omega L)$

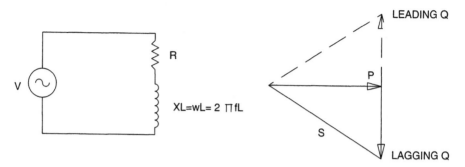

FIGURE 6.5 Lagging and leading reactive power representation.

Multiplying the numerator and the denominator by $(R - j\omega L)$,

$$S = V^2(R - j\omega L)/(R^2 + \omega^2 L^2)$$

Separating the terms,

$$S = V^2R/(R^2 + \omega^2 L^2) - jV^2\omega L/(R^2 + \omega^2 L^2)$$

$$S = P - jQ \qquad (6.6)$$

The $-Q$ indicates that the reactive power is lagging. By supplying a leading reactive power equal to Q, we can correct the power factor to unity.

From Eq. (6.4), $Q/P = \tan\emptyset$. From Eq. (6.5), $Q/P = \omega L/R = \tan\emptyset$ and $\emptyset = \tan^{-1}(\omega L/R)$, thus:

$$\text{Power factor} = \cos\emptyset = \cos(\tan^{-1}\omega L/R) \qquad (6.7)$$

Example: In the circuit shown in Figure 6.5, $V = 480$ V, $R = 1\ \Omega$, and $L = 1$ mH; therefore,

$$X_L = \omega L = 2\pi fL = 2\pi \times 60 \times .001 = 0.377\ \Omega$$

From Eq. (6.6),

Active power = $P = V^2R/(R^2 + \omega^2 L^2) = 201.75$ kW

Reactive power = $Q = V^2\omega L/(R^2 + \omega^2 L^2) = 76.06$ kVAR

Power factor angle = $\emptyset = \tan^{-1}(Q/P) = \tan^{-1}(0.377) = 20.66°$

Power factor = $PF = \cos\emptyset = 0.936$

The leading reactive power necessary to correct the power factor to 1.0 is 76.06 kVAR.

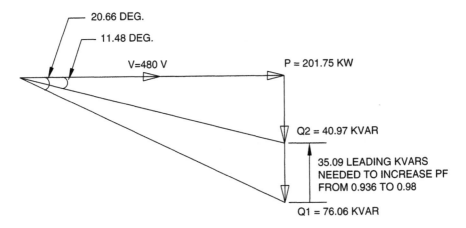

FIGURE 6.6 Power factor correction triangle.

In the same example, what is the leading kVAR required to correct the power factor to 0.98? At 0.98 power factor lag, the lagging kVAR permitted can be calculated from the following:

$$\text{Power factor angle at } 0.98 = 11.48°$$

$$\tan(11.48°) = Q/201.75 = 0.203$$

$$Q = 0.203 \times 201.75 = 40.97 \text{ kVAR}$$

The leading kVAR required in order to correct the power factor to 0.98 = 76.06 – 40.97 = 35.09 (see Figure 6.6).

In a typical power system, power factor calculations, values of resistance, and inductance data are not really available. What is available is total active and reactive power. From this, the kVAR necessary to correct the power factor from a given value to another desired value can be calculated. Figure 6.7 shows the general power factor correction triangles. To solve this triangle, three pieces of information are needed: existing power factor $(\cos\emptyset_1)$, corrected power factor $(\cos\emptyset_2)$, and any one of the following: active power (P), reactive power (Q), or apparent power (S).

- Given P, $\cos\emptyset_1$, and $\cos\emptyset_2$:
  From the above, $Q_1 = P\tan\emptyset_1$ and $Q_2 = P\tan\emptyset_2$. The reactive power required to correct the power factor from $\cos\emptyset_1$ to $\cos\emptyset_2$ is:

$$\Delta Q = P(\tan\emptyset_1 - \tan\emptyset_2)$$

- Given S_1, $\cos\emptyset_1$, and $\cos\emptyset_2$:
  From the above, $Q_1 = S_1\sin\emptyset_1$, $P = S_1\cos\emptyset_1$, and $Q_2 = P\tan\emptyset_2$. The leading reactive power necessary is:

$$\Delta Q = Q_1 - Q_2$$

- Given Q_1, $\cos\emptyset_1$, and $\cos\emptyset_2$:

From the above, $P = Q_1/\tan\emptyset_1$ and $Q_2 = P\tan\emptyset_2$. The leading reactive power necessary is:

$$\Delta Q = Q_1 - Q_2$$

Example: A 5-MVA transformer is loaded to 4.5 MVA at a power factor of 0.82 lag. Calculate the leading kVAR necessary to correct the power factor to 0.95 lag. If the transformer has a rated conductor loss equal to 1.0% of the transformer rating, calculate the energy saved assuming 24-hour operation at the operating load. Figure 6.8 contains the power triangle of the given load and power factor conditions:

Existing power factor angle $= \emptyset_1 = \cos^{-1}(0.82) = 34.9°$

Corrected power factor angle $= \emptyset_2 = \cos^{-1}(0.95) = 18.2°$

$Q_1 = S_1\sin\emptyset_1 = 4.5 \times 0.572 = 2.574$ MVAR

$P = S_1\cos\emptyset_1 = 4.5 \times 0.82 = 3.69$ MW

$Q_2 = P\tan\emptyset_2 = 3.69 \times 0.329 = 1.214$ MVAR

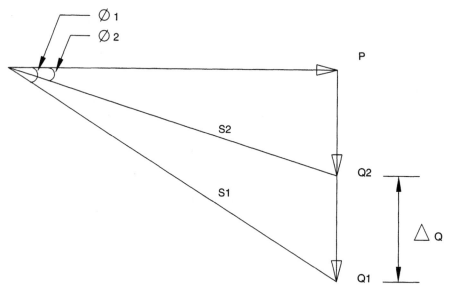

FIGURE 6.7 General power factor correction triangle.

The leading MVAR necessary to improve the power factor from 0.82 to 0.95 = Q_1 − Q_2 = 1.362. For a transformer load with improved power factor S_2:

$$S_2 = \sqrt{(P^2 + Q_2^2)} = 3.885 \text{ MVA}$$

The change in transformer conductor loss = 1.0 $[(4.5/5)^2 - (3.885/5)^2] = 0.206$ p.u. of rated losses, thus the total energy saved = 0.206 × 50 × 24 = 247.2 kWhr/day. At a cost of $0.05/kWhr, the energy saved per year = 247.2 × 365 × 0.05 = $4511.40.

6.6 POWER FACTOR PENALTY

Typically, electrical utilities charge a penalty for power factors below 0.95. The method of calculating the penalty depends on the utility. In some cases, the formula is simple, but in other cases the formula for the power factor penalty can be much more complex. Let's assume that one utility charges a rate of 0.20¢/kVAR–hr for all the reactive energy used if the power factor falls below 0.95. No kVar–hr charges are levied if the power factor is above 0.95.

In the example above, at 0.82 power factor the total kVar–hr of reactive power used per month = 2574 × 24 × 30. The total power factor penalty incurred each month = 2574 × 24 × 30 × 0.20 × 0.01 = $3707. The cost of having a low power factor per year is $44,484. The cost of purchasing and installing power factor correction equipment in this specific case would be about $75,000. It is not difficult to see the cost savings involved by correcting the power factor to prevent utility penalties.

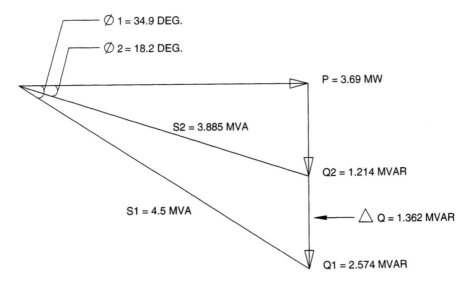

FIGURE 6.8 Power factor triangle for Section 6.4 example.

Another utility calculates the penalty using a different formula. First, kW demand is increased by a factor equal to the 0.95 divided by actual power factor. The difference between this and the actual demand is charged at a rate of \$3.50/kW. In the example, the calculated demand due to low power factor = 3690 × 0.95/0.82 = \$4275, thus the penalty kW = 4275 − 3690 = \$585, and the penalty each month = 585 × \$3.50 = \$2047. In this example, the maximum demand is assumed to be equal to the average demand calculated for the period. The actual demand is typically higher than the average demand. The penalty for having a poor power factor will be correspondingly higher. In the future, as the demand for electrical power continues to grow, the penalty for poor power factors is expected to get worse.

6.7 OTHER ADVANTAGES OF POWER FACTOR CORRECTION

Correcting low power factor has other benefits besides avoiding penalties imposed by the utilities. Other advantages of improving the power factor include:

- Reduced heating in equipment
- Increased equipment life
- Reduction in energy losses and operating costs
- Freeing up available energy
- Reduction of voltage drop in the electrical system

In Figure 6.9, the total apparent power saved due to power factor correction = 4500 − 3885 = 615 kVA, which will be available to supply other plant loads or help minimize capital costs in case of future plant expansion. As current drawn from the source is lowered, the voltage drop in the power system is also reduced. This is important in large industrial facilities or high-rise commercial buildings, which are typically prone to excessive voltage sags.

6.8 VOLTAGE RISE DUE TO CAPACITANCE

When large power factor correction capacitors are present in an electrical system, the flow of capacitive current through the power system impedance can actually

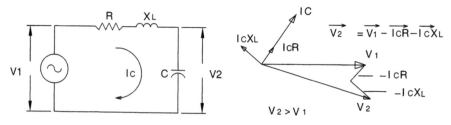

FIGURE 6.9 Schematic and phasor diagram showing voltage rise due to capacitive current flowing through line impedance.

produce a voltage rise, as shown in Figure 6.9. In some instances, utilities will actually switch on large capacitor banks to effect a voltage rise on the power system at the end of long transmission lines. Depending on the voltage levels and the reactive power demand of the loads, the capacitors may be switched in or out in discreet steps. Voltage rise in the power system is one reason why the utilities do not permit large levels of uncompensated leading kVARs to be drawn from the power lines. During the process of selecting capacitor banks for power factor correction, the utilities should be consulted to determine the level of leading kVARs that can be drawn. This is not a concern when the plant or the facility is heavily loaded, because the leading kVARs would be essentially canceled by the lagging reactive power demand of the plant. But, during light load periods, the leading reactive power is not fully compensated and therefore might be objectionable to the utility. For applications where large swings in reactive power requirements are expected, a switched capacitor bank might be worth the investment. Such a unit contains a power factor controller that senses and regulates the power factor by switching blocks of capacitors in and out. Such equipment is more expensive. Figure 6.10 depicts a switched capacitor bank configured to maintain a power factor between two preset limits for various combinations of plant loading conditions.

6.9 APPLICATION OF SYNCHRONOUS CONDENSERS

It was observed in Chapter 4 that capacitor banks must be selected and applied based on power system harmonic studies. This is necessary to eliminate conditions that can actually amplify the harmonics and create conditions that can render the situation considerably worse. One means of providing leading reactive power is by the use of synchronous motors. Synchronous motors applied for power factor control are called synchronous condensers. A synchronous motor normally draws lagging currents, but when its field is overexcited, the motor draws leading reactive currents (Figure 6.11). By adjusting the field currents, the synchronous motor can be made

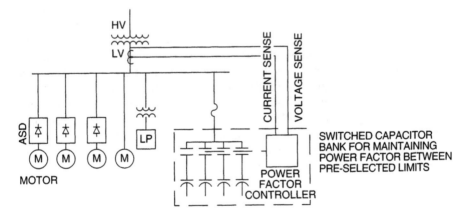

FIGURE 6.10 Schematic of a switched capacitor bank for power factor control between preselected limits for varying plant load conditions.

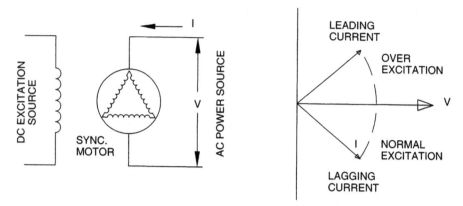

FIGURE 6.11 Synchronous condenser for power factor correction.

to operate in the lagging, unity, or leading power factor region. Facilities that contain large AC motors are best suited for the application. Replacing an AC induction motor with a synchronous motor operating in the leading power factor region is an effective means of power factor control. Synchronous motors are more expensive than conventional induction motors due to their construction complexities and associated control equipment. Some facilities and utilities use unloaded synchronous motors strictly for leading reactive power generation. The advantage of using a synchronous condenser is the lack of harmonic resonance problems sometimes found with the use of passive capacitor banks.

6.10 STATIC VAR COMPENSATORS

Static VAR compensators (SVCs) use static power control devices such as SCRs or IGBTs and switch a bank of capacitors and inductors to generate reactive currents of the required makeup. Reactive power is needed for several reasons. As we saw earlier, leading reactive power is needed to improve the power factor and also to raise the voltage at the end of long power lines. Lagging reactive power is sometimes necessary at the end of long transmission lines to compensate for the voltage rise experienced due to capacitive charging currents of the lines. Uncompensated, such power lines can experience a voltage rise beyond what is acceptable. The reactors installed for such purposes are called line charge compensators.

Static VAR compensators perform both functions as needed. Figure 6.12 contains a typical arrangement of an SVC. By controlling the voltage to the capacitors and inductors, accurate reactive current control is obtained. One drawback of using SVCs is the generation of a considerable amount of harmonic currents that may have to be filtered. The cost of SVCs is also high, so they will not be economical for small power users.

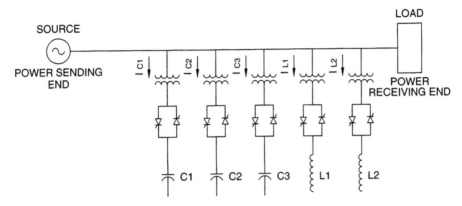

FIGURE 6.12 Static VAR compensator draws optimum amount of leading and lagging currents to maintain required voltage and power factor levels.

6.11 CONCLUSIONS

Good power factor is not necessarily critical for most equipment to function in a normal manner. Having low power factor does not cause a piece of machinery to shut down, but high power factor is important for the overall health of the power system. Operating in a high power factor environment ensures that the power system is functioning efficiently. It also makes economic sense. Electrical power generation, transmission, and distribution equipment have maximum rated currents that the machines can safely handle. If these levels are exceeded, the equipment operates inefficiently and suffers a loss of life expectancy. This is why it is important not to exceed the rated currents for power system equipment. It is also equally important that the available energy production capacity be put to optimum use. Such an approach helps to provide an uninterrupted supply of electrical energy to industries, hospitals, commercial institutions, and our homes. As the demand for electrical energy continues to grow and the resources for producing the energy become less and less available, the idea of not using more than what we need takes on more relevance.

7 Electromagnetic Interference

7.1 INTRODUCTION

Electricity and magnetism are interrelated and exist in a complementary fashion. Any conductor carrying electrical current has an associated magnetic field. A magnetic field can induce voltages or currents in a conductive medium exposed to the field. Altering one changes the other consistent with certain principles of electromagnetic dependency. Electrical circuits are carriers of electricity as well as propagators of magnetic fields. In many pieces of electrical apparatus, the relationship between electrical current and magnetic field is put to productive use. Some examples of utilizing the electromagnetic principle are generators, motors, transformers, induction heating furnaces, electromagnets, and relays, to mention just a few. Everyday lives depend heavily on electromagnetism. In the area of power quality, the useful properties of electromagnetism are not a concern; rather, the interest is in how electromagnetic phenomena affect electrical and electronic devices in an adverse manner. The effect of electromagnetism on sensitive devices is called electromagnetic interference (EMI) and is a rather complex subject. Many voluminous books are available on this subject, and each aspect of EMI is covered in depth in some of them. Here, some of the essential elements of EMI will be discussed in order to give the reader the basic understanding necessary to be able to identify problems relating to this phenomenon.

7.2 FREQUENCY CLASSIFICATION

Table 7.1 shows how the frequency spectrum is classified and primary uses of each type of frequency. While some level of overlapping of frequencies may be found in various books, the frequency classification provided here is generally agreed upon. The mode of interference coupling may not be significantly different for two adjacent frequency bands. But, if the frequency spectrums of interest are two or three bands removed, the interference coupling mode and treatment of the EMI problem could be radically different. One advantage of knowing the frequency bands used by any particular group or agency is that once the offending frequency band is determined (usually by tests), the source of the EMI may be determined with reasonable accuracy. While the problem of EMI is more readily associated with signals in the low-frequency range and beyond, in this book all frequency bands are considered for discussion, as all of the frequency bands — from DC to extremely high frequency (EHF) — can be a source of power quality problems.

7.3 ELECTRICAL FIELDS

Two important properties of electromagnetism — electrical and magnetic fields — will be briefly discussed here; it is not the intent of this book to include an in-depth discussion of the two quantities, but rather to provide an understanding of each phenomenon. An electrical field is present whenever an electrical charge (q) placed in a dielectric or insulating medium experiences a force acting upon it. From this definition, conclude that electrical fields are forces. The field exists whenever a charge differential exists between two points in a medium. The force is proportional to the product of the two charges and inversely proportional to the square of the distance between the two points.

If two charges, q_1 and q_2, are placed at a distance of d meters apart in a dielectric medium of relative permittivity equal to ε_R, the force (F) acting between the two charges is given by Coulomb's law:

$$F = q_1 q_2/\varepsilon_0 \varepsilon_R d^2 \tag{7.1}$$

where ε_0 is the permittivity of free space and is equal to 8.854×10^{-12} F/m. If the medium is free space then ε_R is equal to 1.

Electrical forces may be visualized as lines of force between two points, between which exists a charge differential (Figure 7.1). Two quantities describe the electrical field: electrical field intensity (**E**) and electrical flux density (D). Electrical field intensity is the force experienced by a unit charge placed in the field. A unit charge has an absolute charge equal to 1.602×10^{-19} C; therefore,

$$\mathbf{E} = F/q \tag{7.2}$$

where q is the total charge. The unit of electrical field intensity is volts per meter (V/m) or newtons per coulomb (N/C). Electric field intensity (**E**) is a vector quantity, meaning it has both magnitude and direction (vector quantities are usually described by bold letters and numbers). If q_2 in Eq. (7.1) is a unit charge, then from Eqs. (7.1) and (7.2):

$$\mathbf{E} = q_1/\varepsilon_0 \varepsilon_R d^2 \tag{7.3}$$

Equation (7.3) states that the electric field intensity varies as the square of the distance from the location of the charge. The farther q_1 is located away from q_2, the lower the field intensity experienced at q_2 due to q_1.

Electric flux density is the number of electric lines of flux passing through a unit area. If ψ number of electric flux lines pass through an area A (m^2), then electric flux density is given by:

$$D = \psi/A \tag{7.4}$$

The ratio between the electric flux density and the field intensity is the permittivity of free space or ε_0; therefore,

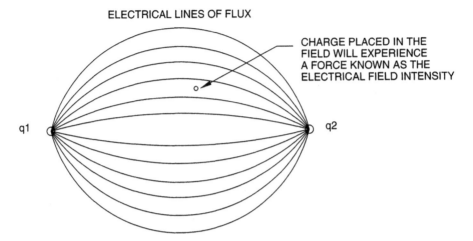

FIGURE 7.1 Electrical lines of flux between two charged bodies.

$$\varepsilon_0 = D/E = 8.854 \times 10^{-12} \text{ F/m}$$

In other mediums besides free space, **E** reduces in proportion to the relative permittivity of the medium. This means that the electric flux density D is independent of the medium.

In power quality studies, we are mainly concerned with the propagation of EMI in space (or air) and as such we are only concerned with the properties applicable to free space. Electrical field intensity is the primary measure of electrical fields applicable to power quality. Most field measuring devices indicate electric fields in the units of volts/meter, and standards and specifications for susceptibility criteria for electrical fields also define the field intensity in volts/meter, which is the unit used in this book.

7.4 MAGNETIC FIELDS

Magnetic fields exist when two poles of the opposite orientation are present: the north pole and the south pole. Two magnetic poles of strengths m_1 and m_2 placed at a distance of d meters apart in a medium of relative permeability equal to μ_R will exert a force (F) on each other given by Coulomb's law, which states:

$$F = m_1 m_2 / \mu_0 \mu_R d^2 \tag{7.5}$$

where μ_0 is the permeability of free space = $4\pi \times 10^{-7}$ **H**/m.

Magnetic field intensity, **H**, is the force experienced by a unit pole placed in a magnetic field; therefore,

$$\mathbf{H} = F/m$$

If m_2 is a unit pole, the field intensity at m_2 due to m_1 is obtained from Eq. (7.5) as:

$$\mathbf{H} = m_1/\mu_O\mu_R d^2 \tag{7.6}$$

Equation (7.6) points out that the magnetic field intensity varies as the square of the distance from the source of the magnetic field. As the distance between m_1 and m_2 increases, the field intensity decreases.

Magnetic fields are associated with the flow of electrical current in a conductor. Permanent magnets are a source of magnetic fields, but in the discussion of electromagnetic fields these are not going to be included as a source of magnetic fields. When current flows in a conductor, magnetic flux lines are established. Unlike electrical fields, which start and terminate between two charges, magnetic flux lines form concentric tubes around the conductor carrying the electrical current (Figure 7.2).

Magnetic flux density (B) is the number of flux lines per unit area of the medium. If Ø number of magnetic lines of flux pass through an area of A (m^2), the flux density $B = Ø/A$. The relationship between the magnetic flux density and the magnetic field intensity is known as the permeability of the magnetic medium, which is indicated by μ. In a linear magnetic medium undistorted by external factors,

$$\mu = \mu_R \times \mu_O \tag{7.7}$$

where μ_O is the permeability of free space and μ_R is the relative permeability of the magnetic medium with respect to free space. In free space, $\mu_R = 1$; therefore, $\mu = \mu_O = 4\pi \times 10^{-7}$ H/m. In a linear medium,

$$\mu = B/H \tag{7.8}$$

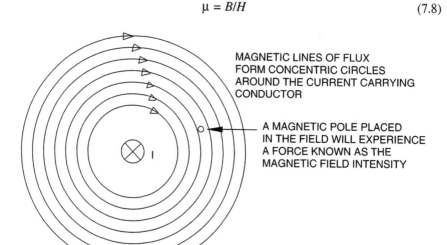

MAGNETIC LINES OF FLUX
FORM CONCENTRIC CIRCLES
AROUND THE CURRENT CARRYING
CONDUCTOR

A MAGNETIC POLE PLACED
IN THE FIELD WILL EXPERIENCE
A FORCE KNOWN AS THE
MAGNETIC FIELD INTENSITY

FIGURE 7.2 Magnetic flux lines due to a current-carrying conductor.

In free space,

$$\mu_O = B/H \qquad (7.9)$$

Magnetic field intensity is expressed in units of ampere-turns per meter, and flux density is expressed in units of tesla (T). One tesla is equal to the flux density when 10^8 lines of flux lines pass through an area of 1 m². A more practical unit for measuring magnetic flux density is a gauss (G), which is equal to one magnetic line of flux passing through an area of 1 cm². In many applications, flux density is expressed in milligauss (mG): 1 mG = 10^{-3} G.

Both electrical and magnetic fields are capable of producing interference in sensitive electrical and electronic devices. The means of interference coupling for each is different. Electrical fields are due to potential or charge difference between two points in a dielectric medium. Magnetic fields (of concern here) are due to the flow of electrical current in a conducting medium. Electrical fields exert a force on any electrical charge (or signal) in its path and tend to alter its amplitude or direction or both. Magnetic fields induce currents in an electrical circuit placed in their path, which can alter the signal level or its phase angle or both. Either of these effects is an unwanted phenomenon that comes under the category of EMI.

7.5 ELECTROMAGNETIC INTERFERENCE TERMINOLOGY

Several terms unique to electromagnetic phenomena and not commonly used in other power quality issues are explained in this section.

7.5.1 Decibel (dB)

The decibel is used to express the ratio between two quantities. The quantities may be voltage, current, or power. For voltages and currents,

$$dB = 20 \log (V_1/V_2) \text{ or } 20 \log (I_1/I_2)$$

where V = voltage and I = current. For ratios involving power,

$$dB = 10 \log (P_1/P_2)$$

where P = active power. For example, if a filter can attenuate a noise of 10 V to a level of 100 mV, then:

$$\text{Voltage attenuation} = V_1/V_2 = 10/0.1 = 100$$

$$\text{Attenuation (dB)} = 20 \log 100 = 40$$

Also, if the power input into an amplifier is 1 W and the power output is 10 W, the power gain (in dB) is equal to 10 log 10 = 10.

7.5.2 RADIATED EMISSION

Radiated emission is a measure of the level of EMI propagated in air by the source. Radiated emission requires a carrier medium such as air or other gases and is usually expressed in volts/meter (V/m) or microvolts per meter (μV/m).

7.5.3 CONDUCTED EMISSION

Conducted emission is a measure of the level of EMI propagated via a conducting medium such as power, signal, or ground wires. Conducted emission is expressed in millivolts (mV) or microvolts (μV).

7.5.4 ATTENUATION

Attenuation is the ratio by which unwanted noise or signal is reduced in amplitude, usually expressed in decibels (dB).

7.5.5 COMMON MODE REJECTION RATIO

The common mode rejection ratio (CMRR) is the ratio (usually expressed in decibels) between the common mode noise at the input of a power handling device and the transverse mode noise at the output of the device. Figure 7.3 illustrates the distinction between the two modes of noise. Common mode noise is typically due to either coupling of propagated noise from an external source or stray ground potentials, and it affects the line and neutral (or return) wires of a circuit equally.

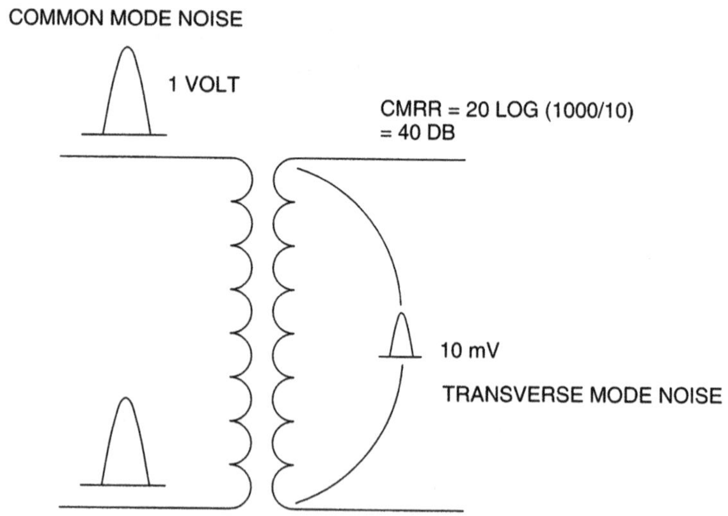

COMMON MODE NOISE

1 VOLT

CMRR = 20 LOG (1000/10)
= 40 DB

10 mV

TRANSVERSE MODE NOISE

FIGURE 7.3 Example of common-mode rejection ratio.

Common mode noise is converted to transverse mode noise in the impedance associated with the lines. Common mode noise when converted to transverse mode noise can be quite troublesome in sensitive, low-power devices. Filters or shielded isolation transformers reduce the amount by which common mode noise is converted to transverse mode noise.

7.5.6 Noise

Electrical noise, or noise, is unwanted electrical signals that produce undesirable effects in the circuits in which they occur.

7.5.7 Common Mode Noise

Common mode noise is present equally and in phase in each current carrying wire with respect to a ground plane or circuit. Common mode noise can be caused by radiated emission from a source of EMI. Common mode noise can also couple from one circuit to another by inductive or capacitive means. Lightning discharges may also produce common mode noise in power wiring,

7.5.8 Transverse Mode Noise

Transverse mode noise is noise present across the power wires to a load. The noise is referenced from one power conductor to another including the neutral wire of a circuit. Figure 7.4 depicts common and transverse mode noises. Transverse mode noise is produced due to power system faults or disturbances produced by other loads. Transverse mode noise can also be due to conversion of common mode noise in power equipment or power lines. Some electrical loads are also known to generate their own transverse mode noise due to their operating peculiarities.

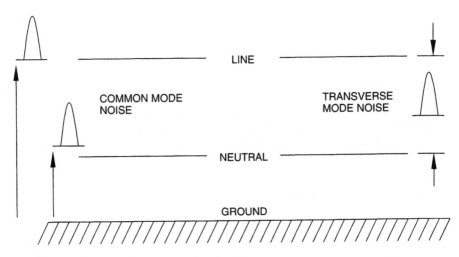

FIGURE 7.4 Common and transverse mode noise.

TABLE 7.1
Frequency Classification

Frequency Classification	Frequency Range	Application
ELF	3–30 Hz	Detection of buried objects
SLF	30–300 Hz	Communication with submarines, electrical power
ULF	300–3000 Hz	Telephone audio range
VLF	3–30 kHz	Navigation, sonar
LF	30–300 kHz	Navigation, radio beacon
MF	300–3000 kHz	AM, maritime radio
HF	3–30 MHz	Shortwave radio, citizen's band
VHF	30–300 MHz	Television, FM, police, mobile
UHF	300–3000 MHz	Radar, television, navigation
SHF	3–30 GHz	Radar, satellite
EHF	30–300 GHz	Radar, space exploration

7.5.9 BANDWIDTH

Bandwidth commonly refers to a range of frequencies. For example, in Table 7.1, a bandwidth of 300 kHz to 300 MHz is assigned to radio broadcast and marine communication. Any filter intended to filter out the noise due to these sources must be designed for this particular bandwidth.

7.5.10 FILTER

A filter consists of passive components such as R, L, and C to divert noise away from susceptible equipment. Filters may be applied at the source of the noise to prevent noise propagation to other loads present in the system. Filters may also be applied at the load to protect a specific piece of equipment. The choice of the type of filter would depend on the location of the noise source, the susceptibility of the equipment, and the presence of more than one noise source.

7.5.11 SHIELDING

A metal enclosure or surface intended to prevent noise from interacting with a susceptible piece of equipment. Shielding may be applied at the source (if the source is known) or at the susceptible equipment. Figure 7.5 illustrates the two modes of shielding.

7.6 POWER FREQUENCY FIELDS

Power frequency fields fall in the category of super low frequency (SLF) fields and are generated by the fundamental power frequency voltage and currents and their harmonics. Because of the low frequency content, these fields do not easily interact with other power, control, or signal circuits. Power frequency electrical fields do not

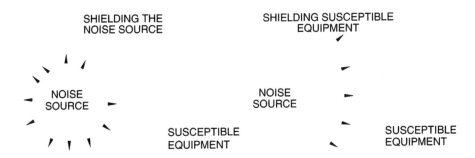

FIGURE 7.5 Radiated noise can be shielded by either shielding the source of noise or by shielding susceptible equipment.

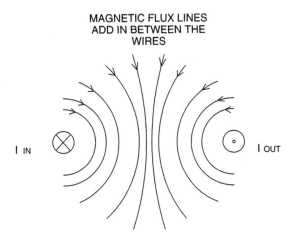

FIGURE 7.6 Magnetic field due to supply and return wires.

easily couple to other circuits through stray capacitance between the circuits. Power frequency magnetic fields tend to be confined to low reluctance paths that consist of ferromagnetic materials. Power frequency currents set up magnetic fields that are free to interact with other electrical circuits and can induce noise voltages at the power frequency.

In a power circuit, magnetic fields caused by the currents in the supply and return wires essentially cancel out outside the space occupied by the wires; however, magnetic fields can exist in the space between the wires (Figure 7.6). Residual electromagnetic force (EMF) attributed to power wiring is rarely a problem if proper wiring methods are used. Typically, power wiring to a piece of equipment is self-contained, with the line, neutral, and ground wires all installed within the same conduit. The net EMF outside the conduit with this arrangement is negligible. Once the power wires enter an enclosure containing sensitive devices, special care should be exercised in the routing of the wires. Figure 7.7 shows the proper and improper ways to route wires within an enclosure. Besides keeping the supply and return wires

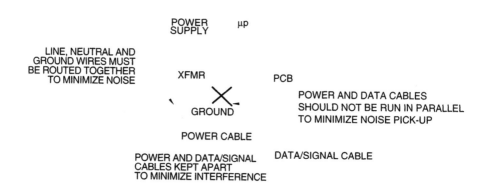

FIGURE 7.7 Equipment wiring to minimize coupling of noise.

in close proximity, it is also important to avoid long parallel runs of power and signal circuits. Such an arrangement is prone to noise pickup by the signal circuit. Also, power and signal circuits should be brought into the enclosure via separate raceways or conduits. These steps help to minimize the possibility of low-frequency noise coupling between the power and the signal circuits.

One problem due to low-frequency electromagnetic fields and observed often in commercial buildings and healthcare facilities is the interaction between the fields and computer video monitors. Such buildings contain electrical vaults, which in some cases are close to areas or rooms containing computer video monitors. The net electromagnetic fields due to the high current bus or cable contained in the vault can interact with computer video monitors and produce severe distortions. The distortions might include ghosting, skewed lines, or images that are unsteady. For personnel that use computers for a large part of the workday, these distortions can be disconcerting. In the high-current electrical vault, it is almost impossible to balance the wiring or bus so that the residual magnetic field is very low. A practical solution is to provide a shielding between the electrical vault and the affected workspaces. The shielding may be in the form of sheets of high conductivity metal such as aluminum. When a low-frequency magnetic field penetrates a high-conductivity material, eddy currents are induced in the material. The eddy currents, which set up magnetic fields that oppose the impinging magnetic field, create a phenomenon called reflection. When a material such as low carbon steel is used for shielding low-frequency magnetic fields, the magnetic fields are absorbed as losses in the ferrous metal. High-permeability material such as Mu-metal is highly effective in shielding low-frequency magnetic fields; however, such metals are very expensive and not very economical for covering large surfaces.

Anomalies in the power wiring are a common cause of stray magnetic fields in commercial buildings and hospitals. Neutral-to-ground connections downstream of the main bonding connection cause some of the neutral current to return via the ground path. This path is not predictable and results in residual magnetic fields due to mismatch in the supply and return currents to the various electrical circuits in the

FIGURE 7.8 Low-frequency electromagnetic field meter used to measure magnetic and electric fields.

facility. While low-frequency electromagnetic fields can interact with computer video monitors or cause hum in radio reception, they do not directly interact with high-speed digital data or communication circuits, which operate at considerably higher frequencies. Figure 7.8 shows how low-frequency electromagnetic fields are measured using an EMF probe, which indicates magnetic fields in milligauss (mG). Magnetic fields as low as 10 mG can interact with a computer video monitor and produce distortion. In typical commercial buildings, low-frequency magnetic fields range between 2 and 5 mG. Levels higher than 10 mG could indicate the presence of electrical rooms or vaults nearby. Higher levels of EMF could also be due to improper wiring practices, as discussed earlier.

7.7 HIGH-FREQUENCY INTERFERENCE

The term EMI is commonly associated with high-frequency noise, which has several possible causes. Figure 7.9 depicts how EMI may be generated and propagated to equipment. Some more common high-frequency EMI sources are radio, television, and microwave communication towers; marine or land communication; atmospheric discharges; radiofrequency heating equipment; adjustable speed drives; fluorescent lighting; and electronic dimmers. These devices produce interference ranging from a few kilohertz to hundreds of megahertz and perhaps higher. Due to their remote

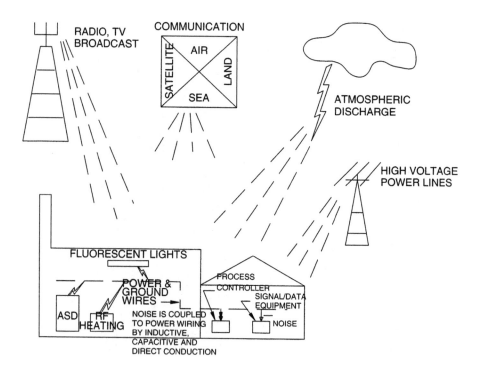

FIGURE 7.9 Common electromagnetic interference (EMI) sources.

distance and because electrical and magnetic fields diminish as the square of the distance from the source, the effects of several of the aforementioned EMI sources are rarely experienced. But, for locations close to the EMI source, the conditions could be serious enough to warrant caution and care. This is why agencies such as the Federal Communications Commission (FCC) have issued maximum limits for radiated and conducted emission for data processing and communication devices using digital information processing. The FCC specifies two categories of devices: class A and class B. Class A devices are intended for use in an industrial or a commercial installation, while class B devices are intended for use in residential environments. Because class B devices are more apt to be installed in close proximity to sensitive equipment, class B limits are more restrictive than class A limits. These standards have to be met by product manufacturers.

It is reasonable to assume that using equipment complying with FCC limits would allow a sensitive device installed next to equipment to function satisfactorily. Unfortunately, this is not always true because internal quirks in the component arrangement or wiring can make a device more sensitive to EMI than a properly designed unit. For example, location and orientation of the ground plane within a device can have a major impact on the equipment functionality. Figure 7.10 indicates the proper and improper ways to provide a ground plane or wire for equipment. In Figure 7.10A, noise coupling is increased due to the large area between the signal

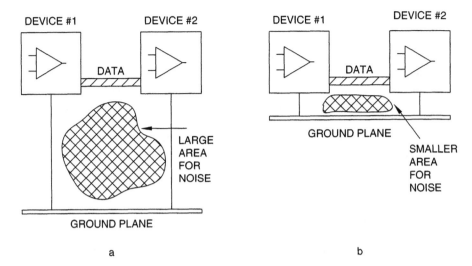

FIGURE 7.10 Location of ground plane or wire can affect noise pickup due to effective ground loop area.

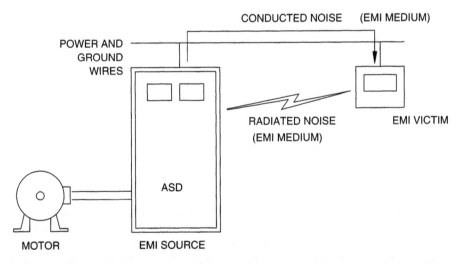

FIGURE 7.11 Criteria for electromagnetic interference (EMI) source, conducting medium, and victim.

and the ground wires. In Figure 7.10B, noise is kept to a minimum by keeping this area small. The same philosophy can be extended to connection of sensitive equipment to power, data, or communication circuits. As much as possible, effective area between the signal wires, between the power wires, and between the wires and the ground should be kept as small as practical.

7.8 ELECTROMAGNETIC INTERFERENCE SUSCEPTIBILITY

To produce electromagnetic interference, three components must exist: (1) a source of interference, (2) a "victim" susceptible to EMI, and (3) a medium for the coupling of EMI between the source and the "victim," which is any device sensitive to the interference. The coupling medium could be inductive or capacitive, radiated through space or transmitted over wires, or a combination of these. Identification of the three elements of EMI as shown in Figure 7.11 allows the EMI to be treated in one of three ways:

- Treatment of the EMI source by isolation, shielding, or application of filters
- Elimination of coupling medium by shielding, use of proper wiring methods, and conductor routing
- Treatment of the "victim" by shielding, application of filters, or location

In some instances, more than one solution may need to be implemented for effective EMI mitigation.

7.9 EMI MITIGATION

7.9.1 SHIELDING FOR RADIATED EMISSION

To control radiated emission, shielding may be applied at the source or at the "victim." Very often it is not practical to shield the source of EMI. Shielding the "victim" involves provision of a continuous metal housing around the device which permits the EMI to be present outside the shield and not within the shield. When the EMI strikes the shield, eddy currents induced in the shield are in a direction that results in field cancellation in the vicinity of the shield. Any device situated within the shield is protected from the EMI. Metals of high conductivity such as copper and aluminum are effective shielding materials in high-frequency EMI applications. In order for the shield to be effective the thickness of the shielding must be greater than the skin depth corresponding to the frequency of the EMI and for the material used as the shield. Table 7.2 provides the skin depths of some typical shielding materials corresponding to frequency. It is evident that for shielding made of copper and aluminum to be effective at low frequencies, considerable metal thickness would be needed. Elimination of air space in the seams of the shielding is very critical to maintaining effectiveness. Special care is necessary when shields are penetrated to allow entry of power or data cables into the shielded enclosure.

7.9.2 FILTERS FOR CONDUCTED EMISSION

Filters are an effective means of providing a certain degree of attenuation of conducted emissions. Filters do not completely eliminate the noise but reduce it to a level that might be tolerated by the susceptible device. Filters use passive components

TABLE 7.2
Skin Depth of Various Materials at Different Frequencies

Frequency	Copper (in.)	Aluminum (in.)	Steel (in.)	Mu-metal (in.)
60 Hz	0.335	0.429	0.034	0.014
100 Hz	0.26	0.333	0.026	0.011
1 kHz	0.082	0.105	0.008	0.003
10 kHz	0.026	0.033	0.003	—
100 kHz	0.008	0.011	0.0008	—
1 MHz	0.003	0.003	0.0003	—
10 MHz	0.0008	0.001	0.0001	—
100 MHz	0.00026	0.0003	0.00008	—
1000 MH	0.00008	0.0001	0.00004	—

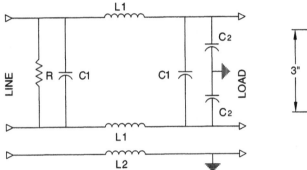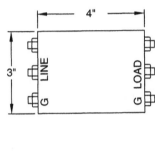

FIGURE 7.12 Typical electromagnetic interference (EMI) filter schematic and outline; the filter yields 60 dB common-mode attenuation and 50 dB transverse mode attenuation between 100 kHz and 1 Mhz.

such as *R*, *L*, and *C* to selectively filter out a certain band of frequencies. A typical passive filter arrangement is shown in Figure 7.12. Passive filters are suitable for filtering a specific frequency band. To filter other bands, a multiband filter or multiple filters are necessary. Filter manufacturers publish frequency vs. attenuation characteristics for each type or model of filter. Prior to application of the filters, it is necessary to determine the range of offending frequencies. Some filter manufacturers will custom engineer and build filters to provide required attenuation at a selected frequency band. For low-level EMI it is sometimes adequate to apply a commercially available filter, which does provide some benefits even though they may be limited. Sometimes filters may be applied in cascade to derive higher attenuation. For instance, two filters each providing 40-dB (100:1) attenuation may be applied in series to derive an attenuation of 80 dB (10,000:1). In reality, the actual attenuation would be less due to parasitic capacitance.

7.9.3 Device Location to Minimize Interference

We saw earlier that electrical and magnetic fields diminish as the square of the distance between the source and the victim. Also, EMI very often is directional. By removing the victim away from the EMI source and by proper orientation, considerable immunity can be obtained. This solution is effective if the relative distance between the source and the victim is small. It is not practical if the source is located far from the victim. For problems involving power frequency EMI this approach is most effective and also most economical.

7.10 CABLE SHIELDING TO MINIMIZE ELECTROMAGNETIC INTERFERENCE

Shielded cables are commonly used for data and signal wires. The configuration of cable shielding and grounding is important to EMI immunity. Even though general guidelines may be provided for shielding cables used for signals or data, each case requires special consideration due to variation in parameters such as cable lengths, noise frequency, signal frequency, and cable termination methodology, each of which can impact the end result. Improperly terminated cable shielding can actually increase noise coupling and make the problem worse. A cable ungrounded at both ends provides no benefits. Generally, shielding at one end also does not increase the attenuation significantly. A cable grounded at both ends, as shown in Figure 7.13, provides reasonable attenuation of the noise; however, with the source and receiver grounded, noise may be coupled to the signal wire when a portion of the signal return current flows through the shields. This current couples to the signal primarily through capacitive means and to a small extent inductively. By using a twisted pair of signal wires, noise coupling can be reduced significantly. As a general rule, it may be necessary to ground the shield at both ends or at multiple points if long lengths are involved. Doing so will reduce the shield impedance to levels low enough to effectively drain any induced noise. At low frequencies, grounding the shield at both ends may not be the best alternative due to the flow of large shield currents. The best shielding for any application is dependent on the application. What is best for one situation may not be the best for a different set of conditions. Sometimes the best solution is determined through actual field experimentation.

7.11 HEALTH CONCERNS OF ELECTROMAGNETIC INTERFERENCE

Electricity and magnetism have been with us since the commercial use of electricity began in the late 1800s, and the demand for electricity has continued to rise since then. Electricity is the primary source of energy at home and at work, and it is not uncommon to see high-voltage transmission lines adjacent to residential areas, which has raised concerns about the effects of electrical and magnetic fields on human health. Engineers, researchers, and physiologists have done considerable work to determine whether any correlation exists between electromagnetic fields and health.

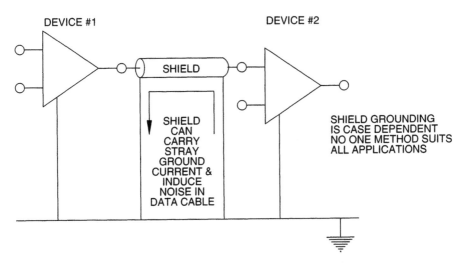

FIGURE 7.13 Cable shield grounding method.

This section provides an overview of the research done in this field so far by the various groups.

Earlier studies on the effects of fields were based on statistical analysis of the incidence of cancer in children and adults who were exposed to electromagnetic fields that were the result of wiring configurations and anomalies found at some of the homes. These studies suggested that the slightly increased risks of cancer in children and adults were due possibly to the fields; however, the risk factors were low. Cancers were reported in homes with slightly higher fields as well as homes with normally expected fields. The number of cases in homes with higher fields was slightly higher, but no overwhelming statistical unbalance between the two scenarios was found.

Later studies involving low-frequency exposure have not clearly demonstrated a correlation between low-level fields and effects on human health. One study observed a slight increase in nervous system tumors for people living within 500 m ($\cong$1600 ft) of overhead power lines, while most recent studies in this field have not found any clear evidence to relate exposure to low-frequency fields with childhood leukemia.

Some experiments on rats and mice show that for continuous exposure at high levels of EMF (400 mG) some physiological changes occur. These EMF levels are well above what humans are normally exposed to at home or at work. One study that exposed humans to high levels of electrical and magnetic fields (greater than 100 times normal) for a short duration found a slowing of heart rate and inhibition of other human response systems.

The studies done so far do not definitively admit or dismiss a correlation between low-frequency magnetic fields and human health. During a typical day, humans are exposed to varying levels of low-frequency electromagnetic fields. This exposure is a byproduct of living in a fast-paced environment. A typical office space will have an ambient low-frequency electromagnetic field ranging between 0.5 and 3 mG.

TABLE 7.3
Low-Frequency Electromagnetic Force Due
to Common Household Equipment

Equipment	EMF 6 in. from Surface (mG)
Personal computer	25
Microwave	75
Range	150
Baseboard heater	40
Electric shaver	20
Hair dryer	150
Television	25

Table 7.3 shows the EMFs produced by some common household electrical appliances. While the EMF levels can be considered high, the exposure duration is low in most cases. It is important to realize that the effects of exposure to low-frequency fields are not clearly known, thus it is prudent to exercise caution and avoid prolonged exposure to electrical and magnetic fields. One way to minimize exposure is to maintain sufficient distance between the EMF source and people in the environment. As we saw earlier, electrical and magnetic fields diminish as the square of the distance from the source. For example, instead of sitting 1 ft away from a table lamp, one can move 2 ft away and reduce EMF exposure to approximately one fourth the level found at 1 ft. It is expected that studies conducted in the future will reveal more about the effects, if any, of low-frequency electromagnetic fields.

7.12 CONCLUSIONS

Electromagnetic fields are all around us and are not necessarily evil. For instance, without these fields radios and televisions would not work, and cell phones would be useless. The garage door opener could not be used from the comfort of a car and the door would not automatically open. Electromagnetic energy is needed for day-to-day lives. It just so happens that some electronic devices may be sensitive to the fields. Fortunately, exposure of such devices to the fields can be reduced. As discussed earlier, shields, filters, and isolation techniques are useful tools that allow us to live in the EMI environment. It is a matter of determining the source of the interference, the tolerance level of the "victim," and the medium that is providing a means of interaction between the two. All EMI problems require knowledge of all three factors for an effective solution.

8 Static Electricity

8.1 INTRODUCTION

The term *static electricity* implies electricity not in motion or electricity that is stationary; in other words, electrons that normally constitute the flow of electricity are in a state of balance appearing static or stationary. In a broad sense, static electricity may represent a battery or a cell, where a potential difference exists across the two terminals. No current can flow until a closed circuit is established between the two, and a fully charged capacitor may be viewed as having static electricity across its plates. However, the aspect of static electricity that this chapter will focus on is the effect of stationary electrical charges produced as the result of contact between two dissimilar materials. Static electricity has been recognized since the mid-1600s. Scientists such as William Gilbert, Robert Boyle, and Otto Von Guericke experimented with static electricity by rubbing together certain materials that had a propensity toward generating electrical charges. Subsequently, large electrostatic generators were built using this principle. Some of the machines were capable of generating electrostatic potentials exceeding 100 kV.

Static electricity is a daily experience. In some instances, the effects are barely noticed, such as when static electricity causes laundry to stick together as it comes out of the dryer. Sometimes static electricity can produce a mild tingling effect. Yet, other static discharges can produce painful sensations of shock accompanied by visible arcs and crackling sounds. Static electricity can be lethal in places such as refineries and grain elevators, where a spark due to static discharge can ignite grain dust or gas vapors and cause an explosion. An understanding of how static electricity develops and how it can be mitigated is essential to preventing problems due to this phenomenon. This chapter discusses static electricity and its importance in the field of electrical power quality.

8.2 TRIBOELECTRICITY

Triboelectricity represents a measure of the tendency for a material to produce static potential buildup. Figure 8.1 contains the triboelectric series for some common materials. The farther apart two materials are in this series, the greater the tendency to generate static voltages when they come into contact with each other. Cotton, due to its ability to absorb moisture, is used as the reference or neutral material. Other materials such as paper and wood are also found at the neutral portion of the triboelectric series. Substances such as nylon, glass, and air are tribo-positive and materials such as polyurethane and Teflon are tribo-negative in the series. Triboelectric substances are able to part with their charges easily. Substances that come into contact with materials positioned away from the neutral part of the series capture electrical charges more readily.

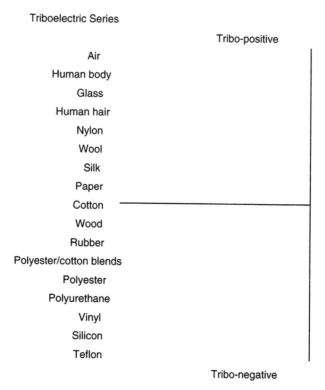

FIGURE 8.1 Triboelectric series of materials. Cotton is used as the neutral or reference material. Tribo-negative materials contain free negative charges, and tribo-positive materials contain free positive charges. These charges are easily transferred to other materials that they might contact.

How are static charges generated? In Figure 8.2, one tribo-negative material (A) and one tribo-positive material (B) come into contact with each other. Figure 8.2A shows the two materials prior to contact and Figure 8.2B illustrates the condition just after contact. During contact, electrons from the negative materials are quickly transferred to the positive material. Some of the electrons neutralize the free positive charges in B; the rest of the electrons remain free and produce a net negative charge. The faster the contact and separation between the substances, the greater the amount of charges trapped on material B. The net charge is a measure of the static electricity. This charge remains on surface B until being discharged to another surface or neutralized.

This is the same phenomenon that takes place when a person walks across the carpet at home and then touches a water faucet or other grounded device in the house. Walking across the carpet allows charges to be picked up from the carpet material, which are stored on the body of the person. When a grounded object such as a faucet is touched, the charge contained on the body is discharged to the ground. At low levels, the discharge produces a tingling sensation. At high charge concentrations, an arc may be produced along with a sharp sensation of pain.

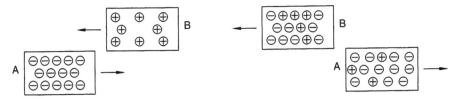

FIGURE 8.2 Mechanism of charge transfer between two materials due to contact and separation.

How many of us can relate to the experience of getting out of a car and receiving an electrical shock when touching the metal body of the car? Walking through a grocery store and experiencing an electrical shock when touching the refrigerated food case is another example of static buildup and discharge. These examples have two things in common: relative motion and contact between two substances that are insulators. A car moving through air, especially if the air is dry and of low humidity, collects electrical charges on its body. When contact is made with the car, electrical charges tend to equalize between the body of the car and the person touching the car. This exchange of charges gives a sensation of electrical shock. The synthetic flooring material used in stores is highly tribo-negative. On a dry day, just walking across the floor can cause accumulation of a charge on a person. When any grounded object is touched, the collected charges are discharged to the object. In these examples, moisture plays an important role in determining the level of electrostatic charges that accumulate on an object or a person. Static discharges are rarely a problem on rainy days due to considerable charge bleeding off that occurs when the air is full of moisture. The moisture may also be present on a person's hands, clothing, and shoes. Also, the body of a car that is damp does not generate large amounts of static electricity. Even though small levels of electrostatic voltages may still be produced on wet days, the levels are not sufficient to cause an appreciable charge buildup and discharge.

8.3 STATIC VOLTAGE BUILDUP CRITERIA

Table 8.1 shows the voltage levels that can build up on a surface due to static electricity. The threshold of perception of static discharge for average humans is between 2000 and 5000 V. A static voltage buildup of 15,000 V or higher is usually required to cause a noisy discharge with accompanying arc. From the table, it can be readily observed that such voltage levels are easily generated during the course of our everyday chores. The type of footwear worn by an individual has an effect on static voltage accumulations. Shoes with leather soles have high enough conductivity to minimize static voltage buildup on the person wearing them. Composite soles and crepe soles have higher resistance, which permits large static buildups. The walking style of a person also affects static discharge. Fast-paced walking on a carpeted floor or synthetic surface tends to produce higher static voltages than slower paced walking, as the electrical charges that are transferred do not have sufficient time to recombine with the opposite-polarity charges present in the material

TABLE 8.1
Static Voltages Generated During Common Day-to-Day Activities

Action	Static Voltage (V)
Person walking across carpet wearing sneakers (50% RH[a])	5000
Plastic comb after combing hair for 5 sec	2000
100% acrylic shirt fresh out of the dryer	20,000
Common grocery store plastic bag (65°F, 50% RH)	300
Car body after driving 10 miles at 60 mph on a dry day (60°F, 55% RH)	4000
Person pushing grocery cart around a store for 5 min (45°F, 40% RH)	10,000

[a] RH = relative humidity.

TABLE 8.2
Electrostatic Susceptibility of Common Semiconductor Devices

Device Type	Susceptibility (V)
MOS/FET	100–200
J-FET	140–10,000
CMOS	250–2000
Schottky diodes, TTL	300–2500
Bipolar transistors	380–7000
ECL	500
SCR	680–1000

they are in contact with. A person's clothing also has an effect. Cotton fabrics do not tend to collect static charges, whereas clothing made of synthetics and polyester allows large static accumulations. A person's skin condition can also influence static discharge. People with drier skin are more prone to large static charge buildup and subsequent discharges that are painful. This is because the surface resistivity of dry skin is considerably higher than for skin that is moist.

To humans, experiencing static discharge may mean nothing more than possible brief discomfort, but its effect on electronic devices can be lethal. Table 8.2 indicates typical susceptibility levels of solid-state devices. Comparing Tables 8.1 and 8.2, it is easy to see how electrostatic voltages are serious concerns in facilities that manufacture or use sensitive electronic devices or circuits. Discharge of electrostatic potential is a quick event, with discharges occurring in a range of between several nanoseconds (10^{-9} sec) and several microseconds (10^{-6} sec). Discharge of static charges over a duration that is too short, causes thermal heating of semiconductors at levels that could cause failures. The reaction times of protective devices are slower than the discharge times of static charges; therefore, static charges are not easily discharged or diverted by the use of protective devices such as surge suppressers or zener diodes.

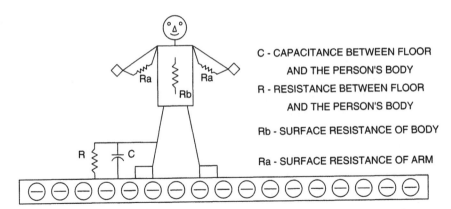

C - CAPACITANCE BETWEEN FLOOR
AND THE PERSON'S BODY

R - RESISTANCE BETWEEN FLOOR
AND THE PERSON'S BODY

Rb - SURFACE RESISTANCE OF BODY

Ra - SURFACE RESISTANCE OF ARM

FIGURE 8.3 Static generator model of person walking across a tribo-negative floor.

Static voltages are not discharged by grounding an insulating surface such as a synthetic carpet or a vinyl floor because electric current cannot flow across the surface or through the body of an insulating medium. This is why control of static charge is a careful science that requires planning. Control of static charge after a facility is completely built is often a difficult and expensive process.

8.4 STATIC MODEL

All models constructed for the study of static voltages involve two dissimilar materials with capacitance coupling formed by an intervening dielectric medium. Figure 8.3 shows an example of a static generator model. Here, two electrodes form capacitor *C*, one of which might be the body of a person and the other highly tribo-negative flooring material. The footwear worn by the person is the dielectric medium helping to form a capacitor-resistive network. The body surface resistance of the person is the discharge path for any accumulated potentials. This resistance determines how quickly the charges might be dissipated through air or via contact with a grounded object. In any problems involving suspected static electricity, the three factors of static generator, capacitor network, and discharge path should be included in the model. Once these are determined, a solution to the problem becomes more evident.

8.5 STATIC CONTROL

In facilities that handle or manufacture sensitive electronic devices, static control is a primary concern. The two important aspects of static control are control of static on personnel and control of static in the facility. Both these issues are part of a composite static control strategy. Static control in personnel starts with attention to the clothing and shoes worn by people working in the environment. Use of cotton clothing is essential, as cotton is neutral in the triboelectric series. Leather-soled shoes are preferred to shoes with composite or crepe soles. Shoe straps made of semiconductive material can be wrapped around a person's ankles and attached to the heels of his shoes so that any charge collecting on the body or clothing of the individual is promptly discharged. Charge accumulation is kept to a minimum in much the same way as a capacitor shunted by a resistor would have smaller charge buildup across its electrodes. Straps worn around the wrist are attached to a ground electrode by means of a suitable ground resistor, which plays an important role in the effectiveness of wrist straps. The resistor helps prevent the buildup of static voltages on the person and limits the rate of discharge of static voltage that does build up to a safe level. Typically, ground resistors in the range of 1 to 2 MΩ are used for the purpose. Too high of a grounding resistance would allow static potential to build up to levels that might be hazardous to sensitive devices being handled. Too low of a resistance could result in a high rate of discharge of static potential, which can cause damage to equipment containing sensitive devices.

Antistatic mats are provided for the operators of sensitive equipment to stand on. Antistatic mats are made of semiconductive material, such as carborized rubber, which provides surface-to-ground resistances ranging between 10^4 and 10^6 Ω. The mats are equipped with pigtail connections for attachment to a ground electrode. When the operator stands on the mat, the semiconductive material discharges any static potential present on the person to a safe level in a methodical manner so as to prevent damage to equipment or electrical shock to the person being discharged. As long as the operator is standing on the mat, static voltages are kept to low levels that will have no deleterious effect on sensitive equipment the person might contact. Figure 8.4 contains a typical representation of an operator in a static discharge environment operating a sensitive electrical machine. As mentioned earlier, the clothing and shoes worn by the person are also part of the overall static control plan and ought to be treated as equally important.

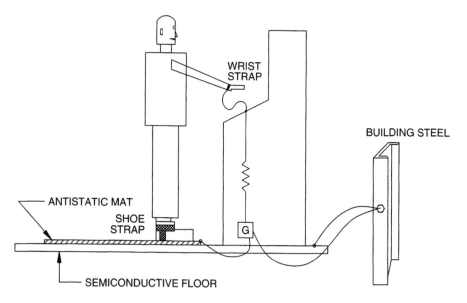

FIGURE 8.4 Static-protective workstation setup showing the use of a wrist strap, a shoe strap, and an antistatic floor mat.

8.6 STATIC CONTROL FLOORS

In static-sensitive areas where no significant level of static voltages may be tolerated, antistatic flooring may be installed. Antistatic flooring is available in two forms: tiles installed on bare concrete surfaces or a coating applied to existing finished floors. Static-control flooring provides surface-to-ground resistances ranging between 10^6 and 10^9 Ω. Semiconductive property enables prompt discharge of static potential accumulated on any person entering the space protected by the floor. Antistatic tiles come in various sizes that can be applied with an adhesive agent to any finished concrete floor surface. Liquid antistatic coatings are applied to clean, finished floor surfaces using any conventional application methods such as rollers or fine brushes. Once cured, such a surface coating provides a semiconductive surface suitable for static prevention. Several precautions are necessary in the installation and care of antistatic floors. Floor mats should be provided at all entrances to the protected area so the amount of debris (dust or dirt) on the floor is kept to a minimum. Floors may be occasionally damp mopped to remove accumulated debris from the floor surface, but floor wax is not to be applied to antistatic floors. Application of floor wax reduces the effectiveness of the flooring in reducing static buildups and in some cases can actually worsen the situation. Grounding the static-control floor is also essential. This is accomplished by using strips of copper in intimate contact with the floor material and bonding the strips to the building ground grid system. Multiple locations of the flooring should be bonded to ground to create an effective antistatic flooring system. Efforts should be made to prohibit abrasive elements such as shoes with hard heels, the wheels of carts, or forklift trucks. Any marking due to such exposure should be promptly removed using wet mops or other suitable cleaners.

8.7 HUMIDITY CONTROL

Humidity is an important factor that helps to minimize static voltage buildups. A 30- to 50-fold reduction of static voltage buildup may be realized by increasing the humidity from 10% to approximately 70 to 80%. A person walking across a carpet and generating 30,000 V at 10% humidity would possibly generate only 600 to 1000 V if the humidity was increased to 80%. The static potential levels might still be high enough to damage sensitive electronic devices, but these are more easily controlled or minimized to below harmless levels. For enclosed spaces containing susceptible devices, humidity enhancement is an effective means of minimizing static voltage accumulations.

8.8 ION COMPENSATION

As noted earlier, static voltages are due to contact between materials that are triboelectric. Such materials have excess charges that are easily imparted to any surface with which they come into contact. If the contact location is well defined, static voltage generation can be minimized by supplying the location with a steady stream of positive and negative ions, which neutralize charges due to triboelectricity. Any

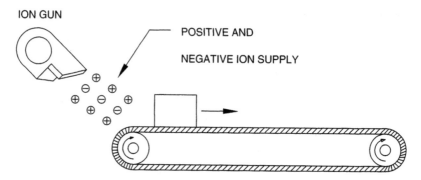

FIGURE 8.5 Use of ion gun to neutralize potential static buildup.

unused ions eventually recombine or discharge to ground. Figure 8.5 illustrates the use of an ion gun in a static control application where friction between a conveyor and the roller generates large static potentials. By providing a steady stream of ions, static potentials can be controlled. Use of an ion gun for static control is suitable only for small spaces. The ions are typically discharged from the gun in the form of a narrow laminar flow with ion concentrations highest at the point of discharge, and the static gun must be pointed directly at the source of the static problem for effective compensation to occur. Away from this targeted location, substantial portions of the positive and negative ions supplied from the gun recombine and are not available for static control. Also, depending upon the application, several ion guns may be necessary to effectively control the static problem.

8.9   STATIC-PREVENTATIVE CASTERS

A problem that has been frequently observed in facilities such as grocery stores is the buildup of static voltage due to the use of metal carts with synthetic casters. Figure 8.6 indicates how static potentials are generated due to relative motion between the cart's wheels and the floor. As the shopper pushes the cart through the store, static voltages are generated at the wheels and transferred to the body of the shopper. Static potentials build to high values in a cumulative manner as the cart is pushed around the store. If the person pushing the cart contacts a grounded object such as the refrigerated food case, sudden discharge of the static potential occurs. Depending on the level of the static voltage, the intensity of the discharge can be high. One means of preventing this phenomenon is the use of antistatic wheels on the carts. These wheels are made of semiconductive materials such as carbon-impregnated rubber or plastics that minimize production of static voltages. Coating the surface of the floor with static-preventive coating is also an option, but due to the degree of traffic involved in applications such as these this is not an effective long-term solution. Incorporating other means of static control, such as the use of antistatic mats at strategic locations of the store, should also be considered. All of these steps should be part of an overall static prevention program for the store.

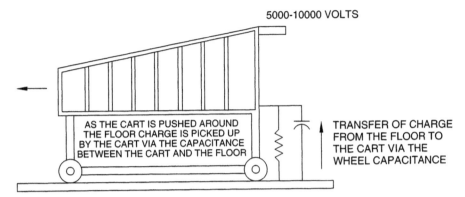

5000-10000 VOLTS

AS THE CART IS PUSHED AROUND THE FLOOR CHARGE IS PICKED UP BY THE CART VIA THE CAPACITANCE BETWEEN THE CART AND THE FLOOR

TRANSFER OF CHARGE FROM THE FLOOR TO THE CART VIA THE WHEEL CAPACITANCE

FIGURE 8.6 Generation of static potential due to movement of the cart wheel on the synthetic floor, which supplies the charges caused by triboelectricity.

8.10 STATIC FLOOR REQUIREMENTS

As discussed earlier, many types of facilities require antistatic flooring to prevent buildup of high static potentials. A healthcare facility is one such example of a building that requires antistatic floors, especially in locations where anesthesia is used and in adjoining spaces. The NFPA 99 Standard for Health Care Facilities makes recommendations for static prevention in such applications. These facilities typically require conductive flooring along with a minimum humidity level of 50%. The requirements for healthcare facilities stipulate a maximum resistance of 10^6 Ω for floor resistance measured (Figure 8.7) by using two test electrodes, each weighing 5 lb with a circular contact area 2.5 inches in diameter. The surface is made of aluminum or tinfoil backed by a 0.25-inch-thick layer of rubber. The electrodes are placed 3 ft apart on the floor to be tested. The resistance between the points is measured with an ohm meter, which has an open circuit voltage of 500 VDC and nominal internal impedance of not less than 100,000 Ω. Usually, five or more measurements are made in each room and the values averaged. No individual measurements should be greater than 5 MΩ, the average value should not be less than 25,000 Ω, and no individual measurements should be less than 10,000 Ω. Measurements should also be made between the flooring and the ground grid system of the room, and these values should also be as specified above. Upper limit is not stipulated for resistance measurements made between one electrode and the ground. The lower limit of 25,000 Ω is intended to limit the current that can flow under fault conditions. Such guidelines may also be adopted for facilities that house sensitive electrical or electronic devices. Typically, measurements are made after installation of a new floor. With use, the resistance values typically increase; therefore, periodic tests are necessary to assess the condition of the floor. If high-resistance locations are found, the floor should be cleaned or retreated as needed to ensure that the floor will continue to provide adequate performance.

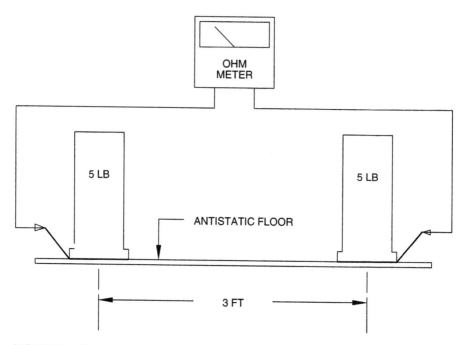

FIGURE 8.7 Measurement of surface resistance using 5-lb electrodes according to the NFPA 99 Standard for Health Care Facilities.

8.11 MEASUREMENT OF STATIC VOLTAGES

Static voltages are measured using an electrostatic meter, a handheld device that utilizes the capacitance in air between a charged surface and the meter membrane. Figure 8.8 shows how a static meter is used to measure static voltages. The meters are battery powered and self-contained; the meter scale is calibrated according to the distance of the meter membrane from the point at which static potentials are to be measured. Static meters are useful for detecting static potentials ranging between 100 and 30,000 V.

8.12 DISCHARGE OF STATIC POTENTIALS

What should be considered a safe static potential level? From Table 8.2, a potential of 100 V may be established as the maximum permissible level for facilities handling or using sensitive devices. A model for safe discharge of static potentials might be developed as follows. A capacitor (C) charged to a voltage of E and discharged through a resistance R will discharge exponentially as determined by the following expression:

$$V = Ee^{-t/RC}$$

where t is the instant in time after closing the switch at which the value of V is required. The voltage across the capacitor decreases exponentially as dictated by the

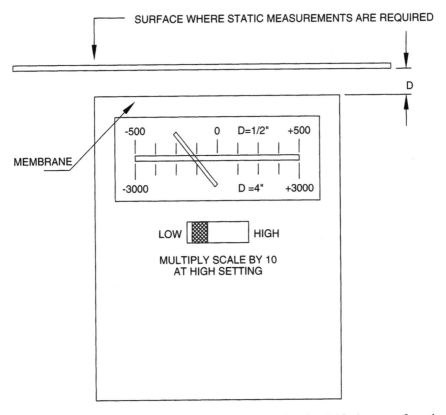

FIGURE 8.8 Static voltmeter. The meter scale is calibrated at .5 and 4 inches away from the test surface.

product of the quantity *RC*, which is known as the time constant of the series resistive/capacitive circuit. The circuit model is shown in Figure 8.9

Example: A triboelectric material with a capacitance of 1 μF is charged to a potential of 20,000 V. What is the value of the resistance required to discharge the material to a safe voltage of 100 V in 1 sec? The expression is given by:

$$100 = 20,000e^{-1.0/R(0.000001)}$$

$$1/e^{1,000,000/R} = 0.005$$

Therefore,

$$e^{1,000,000/R} = 200$$

$$1,000,000/R = \ln(200) = 5.298$$

$$R \cong 189 \ \kappa\Omega$$

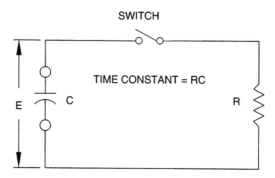

FIGURE 8.9 Capacitance discharge configuration used in static voltage discharge model.

This is the maximum value of resistance to be used to discharge the capacitor to 100 V in 1 sec. In the same example, if the capacitor was initially charged to 30,000 V and using $R = 189$ kΩ, the time to discharge to 100 volts is 1.078 sec (the reader is encouraged to work this out).

In the design of a static-control system, parameters such as capacitance of the personnel, maximum anticipated potential static buildup, and the time to discharge the personnel to safe levels should be known for the model. This is also true when designing static discharge systems for containers entering static protected environments. Such containers should be discharged to safe levels prior to entering the protected space.

8.13 CONCLUSIONS

Static potentials are troublesome in many ways. While examining many different types of facilities experiencing static phenomena, the author has seen firsthand the damaging effects of such static voltage accumulations. In one case, static voltage problems resulted in disruption of operation of a car dealership by locking up the computers several times a day. A semiconductor manufacturing facility was affected due to static potentials building up to levels exceeding 30,000 V. The voltages built up on personnel walking across the production floor on metal gratings that had been coated with a synthetic coating to prevent corrosion. Grocery stores have been prone to static problems primarily due to the use of carts with wheels made of synthetic materials that are highly nonconductive. A facility that handles hazardous chemicals was shut down by the local jurisdiction because static voltages were creating a variety of problems, including malfunction of material-handling equipment. While the underlying problem was the same in each of these cases, the cures were different. In some instances, the problem was corrected by a single fix and in other cases a combination of fixes was necessary. Static electricity is not easy to identify because even at levels far below the threshold of human perception equipment damage or malfunction can result. This chapter has attempted to provide the basic tools necessary to identify static potentials and solutions for dealing with them.

9 Measuring and Solving Power Quality Problems

9.1 INTRODUCTION

Comprehensive knowledge of power quality issues is important in today's electrical power system operating environment, but the ultimate purpose of learning about power quality is to be able to solve power quality problems. Whether the reader is going to put on personal protective equipment and set up instrumentation to determine the problem or entrust someone else to perform this task, information on how to actually accomplish this is vital. Solving power quality problems depends on acquiring meaningful data at the optimum location or locations and within an expedient time frame. In order to acquire useful and relevant data, instruments most suited for a particular application should be utilized. Most power quality problems that go unrecognized are due to use of instruments not ideally suited for that application. One also needs to have a sense about the location or locations where data need to be collected and for how long. After the data is acquired, sort it to determine what information is pertinent to the problem on hand and what is not. This process requires knowledge of the power system and knowledge of the affected equipment. Initially, all data not determined to be directly useful should be set aside for later use. All data deemed to be relevant should be prioritized and analyzed to obtain a solution to the problem. It should be stressed once again that some power quality problems require not a single solution but a combination of solutions to obtain the desired end results. In this chapter, some of the power quality instrumentation commonly used will be discussed and their application in the power quality field will be indicated.

9.2 POWER QUALITY MEASUREMENT DEVICES

9.2.1 HARMONIC ANALYZERS

Harmonic analyzers or harmonic meters are relatively simple instruments for measuring and recording harmonic distortion data. Typically, harmonic analyzers contain a meter with a waveform display screen, voltage leads, and current probes. Some of the analyzers are handheld devices and others are intended for tabletop use. Some instruments provide a snapshot of the waveform and harmonic distortion pertaining to the instant during which the measurement is made. Other instruments are capable of recording snapshots as well as a continuous record of harmonic distortion over time. Obviously, units that provide more information cost more. Depending on the power quality issue, snapshots of the harmonic distortion might suffice. Other problems, however, might require knowledge of how the harmonic distortion characteristics change with plant loading and time.

What is the largest harmonic frequency of interest that should be included in the measurement? It has been the author's experience that measurements to the 25th harmonics are sufficient to indicate the makeup of the waveform. Harmonic analyzers from various manufacturers tend to have different, upper-harmonic-frequency measurement capability. As described in Chapter 4, harmonic distortion levels diminish substantially with the harmonic number. In order to accurately determine the frequency content, the sampling frequency of the measuring instrument must be greater than twice the frequency of the highest harmonic of interest. This rule is called the Nyquist frequency criteria. According to Nyquist criteria, to accurately determine the frequency content of a 60-Hz fundamental frequency waveform up to the 25th harmonic number, the harmonic measuring instrument must have a minimum sampling rate of 3000 (25 × 60 × 2) samples per second. Of course, higher sampling rates more accurately reflect the actual waveform.

Measurement of voltage harmonic data requires leads that can be attached to the points at which the distortion measurements are needed. Typical voltage leads are 4 to 6 ft long. At these lengths, cable inductance and capacitance are not a concern, as the highest frequency of interest is in the range of 1500 to 3000 Hz (25th to 50th harmonic); therefore, no significant attenuation or distortion should be introduced by the leads in the voltage distortion data.

Measuring current harmonic distortion data requires some special considerations. Most current probes use an iron core transformer designed to fit around the conductors in which harmonic measurements are needed (Figure 9.1). Iron-core current probes are susceptible to increased error at high frequencies and saturation at currents higher than the rated values. Prior to installing current probes for harmonic distortion tests, it is necessary to ensure that the probe is suitable for use at high frequencies without a significant loss in accuracy. Manufacturers provide data as to the usable frequency range for the current probes. The probe shown in Figure 9.1 is useful between the frequencies of 5 Hz and 10 kHz for a maximum current rating of 500 A RMS. It should be understood that, even though the probe might be rated for use at the higher frequencies, there is an accompanying loss of accuracy in the data. The aim is to keep the loss of accuracy as low as possible. At higher frequencies, currents and distortions normally looked at are considerably lower than at the lower frequencies, and some loss of accuracy at higher frequencies might not be all that important. Typically, a 5.0% loss in accuracy might be expected, if the waveform contains significant levels of higher order harmonics.

Figure 9.2 shows the use of a handheld harmonic measuring instrument. This particular instrument is a single-phase measurement device capable of being used in circuits of up to 600 VAC. Table 9.1 provides a printout of harmonic distortion data measured at a power distribution panel supplying a small office building. The table shows the voltage and current harmonic information to the 31st harmonic frequency. Along with harmonic distortion, the relative phase angle between the harmonics and the fundamental voltage is also given. Phase angle information is useful is assessing the direction of the harmonic flow and the location of the source of the harmonics.

FIGURE 9.1 Current probe for measuring currents with waveform distortion due to harmonics.

A point worth noting is that the harmonics are shown as a percent of the total RMS value. IEEE convention presents the harmonics as a percent of the fundamental component. Using the IEEE convention would result in higher harmonic percent values. As pointed out in Chapter 4, it does not really matter what convention is used as long as the same convention is maintained throughout the discussion.

Figure 9.3 shows a tabletop harmonic analyzer for measuring harmonic distortion snapshots and harmonic distortion history data for a specified duration. Table 9.2 contains the harmonic current distortion snapshot data recorded at a lighting panel in a high-rise building. Figure 9.4 provides the current waveform and a record of the current history at the panel over 5 days. The harmonic distortion snapshots along with the history graph are very useful in determining the nature of the harmonics and their occurrence pattern.

9.2.2 TRANSIENT-DISTURBANCE ANALYZERS

Transient-disturbance analyzers are advanced data acquisition devices for capturing, storing, and presenting short-duration, subcycle power system disturbances. As one might expect, the sampling rates for these instruments are high. It is not untypical for transient-disturbance recorders to have sampling rates in the range of 2 to 4 million samples per second. Higher sampling rates provide greater accuracy in describing transient events in terms of their amplitude and frequency content. Both

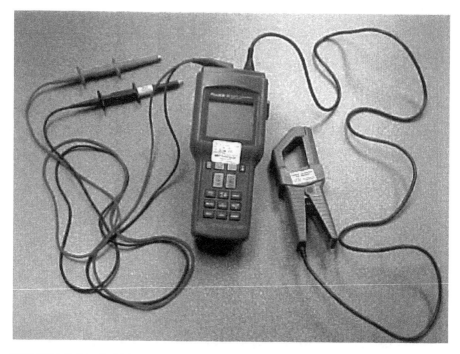

FIGURE 9.2 Handheld harmonic analyzer showing voltage leads and current probe for voltage and current harmonic measurements. (Photograph courtesy of Fluke.)

these attributes are essential for performing transient analysis. The amplitude of the waveform provides information about the potential for damage to the affected equipment. The frequency content informs us as to how the events may couple to other circuits and how they might be mitigated. Figure 9.5 shows a transient that reached peak amplitude of 562 V with a frequency content of approximately 200 kHz. Once such information is determined, equipment susceptibility should be determined. For instance, a 200-V peak impulse applied to a 480-V motor might not have any effect on the motor life; however, the same impulse applied to a process controller could produce immediate failure. Equipment that contains power supplies or capacitor filter circuits is especially susceptible to fast rise-time transients with high-frequency content.

When measuring fast rise time or higher frequency transients, the length of the wires used to connect the instrumentation to the test points becomes very important. In all of these measurements, the leads should be kept as short as possible. Typically, lead lengths of 6 ft or less should not introduce significant errors in the measurements of fast transients. At higher frequencies, cable inductance as well as capacitance become important factors. The use of longer cable lengths in transient measurements results in higher inductance and capacitance and greater attenuation of the transient waveform. Also, in order to minimize noise pickup from external sources, the voltage leads should be kept away from high-voltage and high-current conductors, welding equipment, motors, and transformers. The leads should be kept as straight as possible

TABLE 9.1
Voltage and Current Harmonic Spectrum at an Office Building

Harmonics	Frequency	V Magnitude	% V RMS	V (Phase)	I Magnitude	% I RMS	I (Phase)
DC	0	0.09	0.07	0	0.06	0.14	0
1	59.91	122.84	99.82	0	43.44	97.17	-18
2	119.82	0.09	0.07	74	0.11	0.24	-63
3	179.73	6.33	5.14	42	7.63	17.07	150
4	239.64	0.06	0.05	135	0.09	0.21	90
5	299.56	0.2	0.17	-67	6.3	14.09	-49
6	359.47	0.03	0.03	-156	0.05	0.11	-70
7	419.38	1.15	0.93	20	2.22	4.96	116
8	479.29	0.05	0.04	-80	0	0	105
9	539.2	0.91	0.74	108	0.34	0.77	-150
10	599.11	0.02	0.02	-5	0.01	0.01	45
11	659.02	0.42	0.34	-160	0.81	1.8	-56
12	718.93	0.04	0.03	82	0.01	0.03	165
13	778.84	0.13	0.11	-80	0.52	1.16	96
14	838.76	0.02	0.01	60	0.03	0.07	-88
15	898.67	0.66	0.53	84	0.13	0.28	-159
16	958.58	0.01	0.01	-174	0.01	0.03	63
17	1018.49	0.28	0.23	120	0.37	0.82	-124
18	1078.4	0.01	0.01	-146	0.02	0.04	-129
19	1138.31	0.05	0.04	-145	0.33	0.74	52
20	1198.22	0.01	0.01	-13	0.01	0.03	124
21	1258.13	0.17	0.14	44	0.14	0.31	-179
22	1318.04	0.02	0.01	36	0.02	0.04	-90
23	1377.96	0.15	0.12	101	0.08	0.18	-157
24	1437.87	0.02	0.02	-162	0.01	0.03	-156
25	1497.78	0.02	0.02	167	0.12	0.27	-9
26	1557.69	0.04	0.03	-169	0.01	0.03	-86
27	1617.6	0.09	0.07	-32	0.1	0.22	146
28	1677.51	0.02	0.02	-62	0.04	0.1	34
29	1737.42	0.04	0.03	-29	0.03	0.07	70
30	1797.33	0.02	0.02	-33	0.01	0.03	4
31	1857.24	0.04	0.03	-80	0.08	0.18	-39

Note: The table shows the harmonic number, harmonic frequency, magnitudes, percent harmonic in terms of the total RMS, and the phase angle of each with respect to the fundamental voltage.

without sharp bends or loops. In any case, excess lead length should never be wound into a coil.

Current transformers used in transient current measurements must have a peak current rating at least equal to the maximum expected currents; otherwise, current peaks are lost in the data due to saturation of the current probe. Figure 9.6 indicates how current probe saturation resulted in a flat-top current waveform and loss of vital information, making power quality analysis more difficult.

TABLE 9.2
Current Harmonic Spectrum for a Lighting Panel Supplying
Fluorescent Lighting[a]

Harmonics	RMS Value	Phase	% of Fundamental
0	10.298	180	74.603
1	13.804	157.645	100
2	0.209	337.166	1.511
3	2.014	62.148	14.588
4	0.136	333.435	0.983
5	1.187	81.18	8.603
6	0.051	0	0.366
7	0.372	45	2.695
8	0.121	270	0.879
9	0.551	20.433	3.989
10	0.087	324.462	0.63
11	0.272	15.068	1.973
12	0.101	143.13	0.733
13	0.285	6.116	2.064
14	0.083	345.964	0.604
15	0.083	75.964	0.604
16	0.042	284.036	0.302
17	0.243	45	1.762
18	0.054	21.801	0.395
19	0.051	53.13	0.366
20	0.103	348.69	0.747
21	0.04	0	0.293
22	0.103	281.31	0.747
23	0.036	123.69	0.264
24	0.103	101.31	0.747
25	0.02	90	0.147
26	0.062	189.462	0.446
27	0.068	206.565	0.492
28	0.052	78.69	0.374
29	0.187	49.399	1.351
30	0.03	270	0.22
31	0.145	155.225	1.049
32	0.059	210.964	0.427
33	0.113	10.305	0.819
34	0.074	285.945	0.534
35	0.045	26.565	0.328
36	0.096	341.565	0.695
37	0.136	318.013	0.986
38	0.074	254.055	0.534
39	0.109	201.801	0.789
40	0.109	248.199	0.789
41	0.051	143.13	0.366
42	0.103	281.31	0.747

TABLE 9.2 (CONTINUED)
Current Harmonic Spectrum for a Lighting Panel Supplying
Fluorescent Lighting[a]

Harmonics	RMS Value	Phase	% of Fundamental
43	0.103	101.31	0.747
44	0.045	153.435	0.328
45	0.082	330.255	0.591
46	0.136	228.013	0.986
47	0.043	45	0.311
48	0.152	53.13	1.099
49	0.064	251.565	0.463
50	0.02	270	0.147
51	0.133	278.746	0.964
52	0.086	315	0.622
53	0.125	345.964	0.906
54	0.132	274.399	0.956
55	0.032	341.565	0.232
56	0.045	116.565	0.328
57	0.162	90	1.173
58	0.136	131.987	0.986
59	0.064	288.435	0.463
60	0.154	336.801	1.116
61	0.165	47.49	1.193
62	0.122	221.634	0.882
63	0.051	143.13	0.366

Note: Total harmonic distortion = 18.7%.

[a] Phase A current harmonics, June 27, 2001, 08:57:27.

9.2.3 OSCILLOSCOPES

Oscilloscopes are useful for measuring repetitive high-frequency waveforms or waveforms containing superimposed high-frequency noise on power and control circuits. Oscilloscopes have sampling rates far higher than transient-disturbance analyzers. Oscilloscopes with sampling rates of several hundred million samples per second are common. This allows the instrument to accurately record recurring noise and high-frequency waveforms. Figure 9.7 shows the pulse-width-modulated waveform of the voltage input to an adjustable speed AC motor. Such data are not within the capabilities of harmonic analyzers and transient-disturbance recorders. Digital storage oscilloscopes have the ability to capture and store waveform data for later use. Using multiple-channel, digital storage oscilloscopes, more than one electrical parameter may be viewed and stored. Figure 9.8 shows the noise in the ground grid of a microchip manufacturing facility that could not be detected using other instrumentation. The noise in the ground circuit was responsible for production shutdown at this facility.

FIGURE 9.3 Three-phase harmonic and disturbance analyzer for measuring voltage and current harmonics, voltage and current history over a period of time, voltage transients, and power, power factor, and demand. (Photograph courtesy of Reliable Power Meters.)

Selection of voltage probes is essential for proper use of oscilloscopes. Voltage probes for oscilloscopes are broadly classified into passive probes and active probes. Passive probes use passive components (resistance and capacitance) to provide the necessary filtering and scale factors necessary. Passive probes are typically for use in circuits up to 300 VAC. Higher voltage passive probes can be used in circuits of up to 1000 VAC. Most passive probes are designed to measure voltages with respect to ground. Passive probes, where the probe is isolated from the ground, are useful for making measurements when connection to the ground is to be avoided. Active probes use active components such as field effect transistors to provide high input impedance to the measurements. High input impedance is essential for measuring low-level signals to minimize the possibility of signal attenuation due to loading by the probe itself. Active probes are more expensive than passive probes. The high-frequency current probe is an important accessory for troubleshooting problems

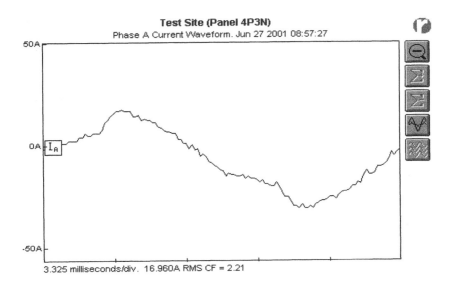

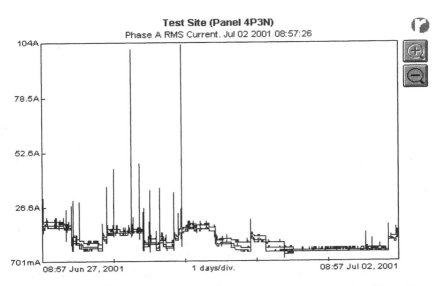

FIGURE 9.4 Current waveform and current history graph at a lighting panel supplying fluorescent lighting.

using an oscilloscope. By using the current probe, stray noise and ground loop currents in the ground grid can be detected.

9.2.4 DATA LOGGERS AND CHART RECORDERS

Data loggers and chart recorders are sometimes used to record voltage, current, demand, and temperature data in electrical power systems. Data loggers and chart

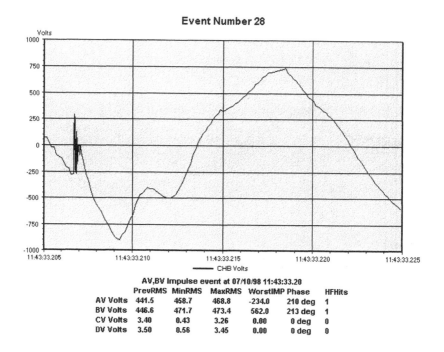

FIGURE 9.5 Switching transient disturbance with a peak of 562 V and a frequency content of 20 kHz.

recorders are slow-response devices that are useful for measuring steady-state data over a long period of time. These devices record one sample of the event at predetermined duration, such as 1 sec, 2 sec, 5 sec, etc. The data are normally stored within the loggers until they are retrieved for analysis. The data from data loggers and chart recorders are sufficient for determining variation of the voltage or current at a particular location over an extended period and if there is no need to determine instantaneous changes in the values. In some applications, this information is all that is needed. But, in power quality assessments involving transient conditions, these devices are not suitable. The advantage of data loggers is that they are relatively inexpensive compared to power quality recording instrumentation. They are also easier to set up and easier to use. The data from the device may be presented in a text format or in a graphical format. Figure 9.9 is the recording of current data at the output of a power transformer using a data logger. The data were produced at the rate one sample every 5 sec. Data loggers and chart recorders are not intended for installation directly on power lines. They are designed to operate using the low-level output of suitable voltage, current, or temperature transducers; however, care should be exercised in the installation and routing of the wires from the transducers to ensure that the output of the transducers is not compromised due to stray noise pickup. Also, data loggers and chart recorders do not provide information about the waveshape of the measured quantity. If that level of information is needed, a power quality analyzer should be used instead.

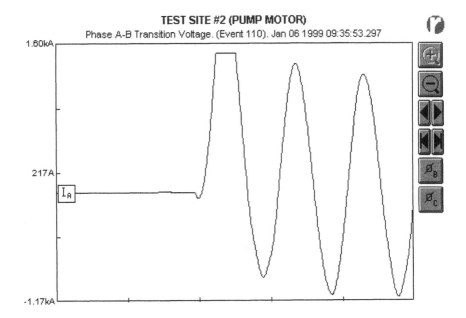

FIGURE 9.6 Current transformer saturation resulting in the loss of vital peak current information.

9.2.5 TRUE RMS METERS

The term *true RMS* is commonly used in power quality applications. What are true RMS meters? As we saw in previous chapters, the RMS value of the current or voltage can be substantially different from the fundamental component of the voltage or current. Using a meter that measures average or peak value of a quantity can produce erroneous results if we need the RMS value of the waveform. For waveforms rich in harmonics, the average and peak values would be considerably different than waveforms that are purely sinusoidal or close to sinusoidal. Measuring the average or peak value of a signal and scaling the values to derive a RMS value would lead to errors.

For example, consider a square wave of current as shown in Figure 9.10. The average and peak reading meters indicate values of 111 A and 70.7 A RMS current, respectively. The square waveform has an average value of 100 A. The peak value of the waveform also has a value of 100 A. In order to arrive at the RMS value, the 100 A average value is multiplied by 1.11, The ratio between the RMS and the average value of a pure sinusoidal waveform is 1.11. The peak reading meter would read the 100 A peak value and multiply it by 0.707 to arrive at the RMS value of 70.7 A, with 0.707 being the ratio between the RMS value and the peak value of a pure sinusoid waveform. The disparities in the values are quite apparent. Figure 9.10 also shows a triangular waveform and the corresponding current data that would be reported by each of the measuring instruments.

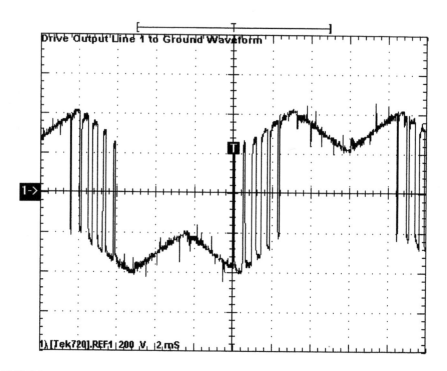

FIGURE 9.7 Pulse-width-modulated waveform from an adjustable speed drive output.

Analog panel meters give erroneous readings when measuring nonsinusoidal currents. Due to higher frequency components, analog meters tend to indicate values that are lower than the actual values. The presence of voltage and current transformers in the metering circuit also introduces additional errors in the measurements.

True RMS meters overcome these problems by deriving the heating effect of the waveform to produce an accurate RMS value indication. After all, RMS value represents the heating effect of a voltage or current signal. Most true RMS meters do not provide any indication of the waveform of the quantity being measured. To accomplish this, the meters require high-frequency signal sampling capability. The sampling rate should satisfy Nyquist criteria in order to produce reasonably accurate results. Some benchtop RMS meters do have the sampling capability and ports to send the information to a computer for waveform display.

9.3 POWER QUALITY MEASUREMENTS

The first step in solving power quality problems is to determine the test location or locations. Even the best available power quality instrumentation is only as good as the personnel using it. Setting up instrumentation at a location that is not optimum with respect to the affected equipment can yield misleading or insufficient information. Electrical transients are especially prone to errors depending on the type of the instrument used and its location. The following example might help to make this point clear.

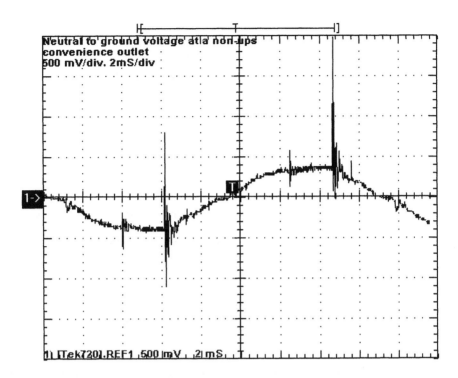

FIGURE 9.8 Electrical noise in the ground grid of a computer center at a microchip manufacturing plant.

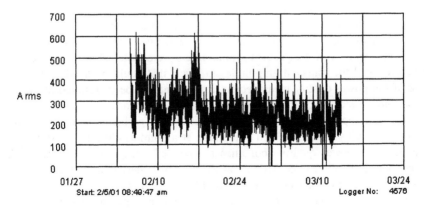

FIGURE 9.9 Current data from a data logger for one month of tests.

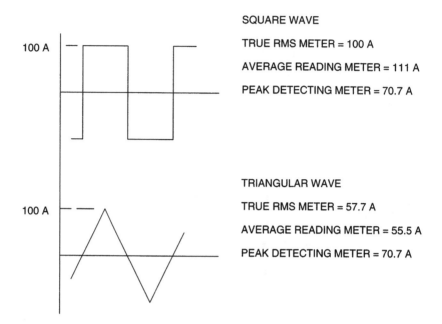

FIGURE 9.10 Variation in rms measurements when using different types of meters.

Example: A large mainframe printing machine was shutting down randomly with no apparent cause. The machine was installed in a computer data center environment and was supplied from an uninterruptible power source (UPS) located about 10 ft from the machine. The power cord from the UPS to the printer was a 15-ft, three-conductor cable. Simultaneous measurement of power quality at the printer input terminals and the UPS terminals supplying the printer revealed that, while transients were present at the machine, no corresponding transients were evident at the UPS. In this case, the 15 ft of cable was sufficient to mask the transient activity. It was determined that the transients were caused by the printer itself due to its large current inrush requirement during the course of printing. The printer contained sensitive voltage detection circuitry which was causing the printer to shut down. To take care of the problem, inline filters were installed at the printer input which reduced the transient amplitudes to levels that could be lived with. In this case, if the power quality measurement instrument had been installed at the UPS output only, the cause of the problem would have gone undetected.

The best approach to investigating power quality problems is to first examine the power quality to the affected equipment at a point as close as possible to the equipment. If power quality anomalies are noticed, then move upstream with a process-of-elimination plan. That is, at each location determine if the problem is due to load-side anomalies or line-side problems. Even though this process is time consuming and perhaps cost ineffective, valuable information can be obtained. Understanding and solving power quality problems is rarely quick and easy.

9.4 NUMBER OF TEST LOCATIONS

If at all possible, power quality tests should be conducted at multiple locations simultaneously. The data obtained by such means are useful in determining the nature of the power quality problem and its possible source as quickly as possible. If simultaneous monitoring is not feasible due to cost or other factors, each location may be individually monitored, taking care to ensure similar operating environments for testing at each location to allow direct comparison of information. The number of test locations would depend on the nature of the problem and the nature of the affected equipment. For example, in Figure 9.11, if power quality problems are observed at location C and not at B, it is not necessary to monitor A. On the other hand, if problems are noticed at C and B, then location A should be tested as well as location D, if necessary. The experience of the power quality engineer becomes important when trying to resolve the different scenarios. For a large facility with multiple transient sources and susceptible equipment, the challenge can be immense.

9.5 TEST DURATION

Deciding upon the length of time that each test point should be monitored for adequate data collection is something that even a well-trained engineer has to struggle with at times. Ideally, you would want to continue tests until the actual cause shows up. This is not always feasible, and the approach can be quite costly. In power quality tests, you are looking not only for the actual failure mode to repeat itself (which would be ideal) but also for any event that might show a tendency toward failure.

Example: The main circuit breaker for a large outlet store would trip randomly. Each occurrence was accompanied by a loss of revenue, not to mention customer dissatisfaction. To determine the cause, the three-phase currents as well as the sum of the phase and the neutral currents were monitored for 3 days. Even though no trips were produced during the time, appreciable phase-neutral residual current was noticed. The sum of the phase and the neutral currents should be zero except in case of a phase-to-ground fault. The problem was traced to a power distribution panel where the neutral bus was also bonded to ground, resulting in residual currents at the main switchboard. Under certain loading conditions, the residual current was high enough to cause the main circuit breaker to trip in a groundfault mode. Once this situation was corrected, the facility experienced no more breaker trips at the main. This is an example of how a particular power quality problem can lead to a solution even when the failure mode cannot be repeated during testing.

Another problem encountered by the author involved trips on a solid-state motor starter, which indicated that the problem was due to a ground fault. The circuit was monitored for a month before any indication of stray ground currents (Figure 9.12) was noticed. If the tests had been discontinued after a week or so, the cause of the trips would have gone undetected, which again emphasizes how difficult it can be to solve power quality problems.

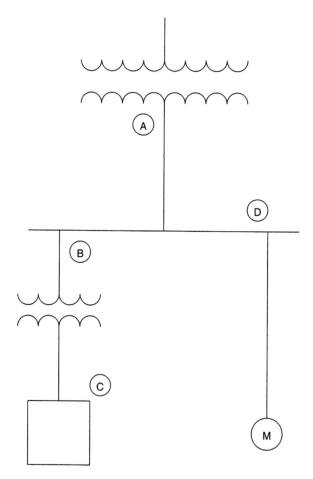

FIGURE 9.11 Test locations for power quality instrumentation.

As a general rule, it is necessary to test each location for at least one week unless results definitely indicate power quality issues at the location that could be causing problems. In such a case, the interval could be shortened. Most power quality issues or tendencies present themselves within this time frame. The actual test durations depend on the experience of the power quality engineers and their comfort level for deriving conclusions based on the data produced. Test duration may be shortened if different power system operating conditions that can cause power system disturbances can be created to generate an adequate amount of data for a solution. Once again, an experienced power quality engineer can help in this process. It is also important to point out that using power quality tendencies to generate conclusions can be risky. This is because under certain conditions more than one power quality problem can produce the same type of symptoms, in which case all possible scenarios should be examined.

Example: A solid-state motor starter was tripping during startup of the motor. Power quality measurements indicated large current draw during the startup. The

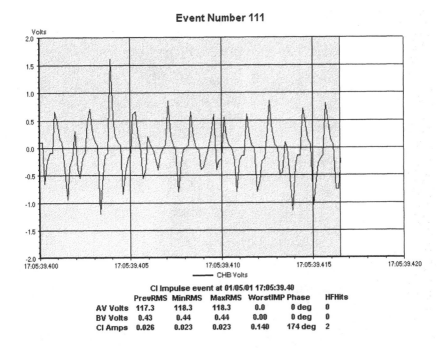

FIGURE 9.12 Stray ground current at the output of a motor that caused the adjustable speed drive to shut down. This event was not recorded until a month after the start of the test.

trips were thought to be due to the starting currents, which exceeded the setting of the starter protection. The actual cause, however, was severe undervoltage conditions produced during startup. The source feeding the starter was not a rigid circuit, causing a large voltage drop during motor start. The excessive current draw was due to severe undervoltage conditions. Once the source to the adjustable speed drive was made rigid, the problem disappeared. In this example, measuring only the current input to the adjustable speed drive would have led to inaccurate conclusions.

9.6 INSTRUMENT SETUP

Setting up instruments to collect power quality data is probably the most critical aspect of testing and also one that most often can decide the end results. Setting up is a time when utmost care must be exercised. The first step is making sure to observe certain safety rules. In the majority of cases, power to electrical equipment cannot be turned off to allow for instrument setup. The facility users want as few interruptions as possible, preferably none. Opening the covers of electrical switch-boards and distribution panels requires diligence and patience. Personal protective equipment (PPE) is an important component of power quality testing. Minimum PPE should contain electrical gloves, safety glasses, and fire-retardant clothing. While removing panel covers and setting up instrument probes it is important to have someone else present in the room. The second person may not be trained in

FIGURE 9.13 Proper personal protective equipment (PPE), which is essential to performing power quality instrument setup and testing. The photograph shows the use of fire-retardant clothing, safety hat, and shoes. Safety glasses must be worn while connecting instrument probes to the test point. The test location shown here is properly barricaded to prevent unauthorized persons from entering the area.

power quality but should have some background in electricity and the hazards associated with it. Figure 9.13 demonstrates the proper use of PPE for performing power quality work.

9.7 INSTRUMENT SETUP GUIDELINES

Installing power quality instruments and probes requires special care. It is preferred that voltage and current probe leads do not run in close proximity to high-current cables or bus, especially if they are subjected to large current inrush. This can inductively induce voltages in the leads of the probes and cause erroneous data to be displayed. Voltage and current lead runs parallel to high-current bus or cable should be avoided or minimized to reduce the possibility of noise pickup. When connecting voltage probes, the connections must be secure. Loose connections are prone to intermittent contact, which can produce false indications of power quality problems. Voltage and current probe leads should be periodically inspected. Leads with damaged insulation or those that look suspect must be promptly replaced to avoid dangerous conditions. While making current measurements, one of the main causes of errors is improper closing of the jaws of the probe. Substantial errors in current measurements and phase angles can be produced due to air gaps across the jaws of the current probes.

It is important to keep the test location well guarded and secured to prevent unauthorized access. The test locations must be secured with barrier tapes or other means to warn people of the hazards. If power distribution panels or switchboards are monitored, all openings created as the result of instrument setup should be sealed to prevent entry by rodents and other pests. All these steps are necessary to ensure that the tests will be completed without accidents.

9.8 CONCLUSIONS

Measurement of power quality requires the use of proper instrumentation to suit the application. The user of the instrument must be well trained in the use and care of the instrumentation. The engineer should be knowledgeable in the field of power quality. Most importantly, the engineer should be safety conscious. All these factors are equally important in solving power quality problems. As indicated earlier, power quality work requires patience, diligence, and perseverance. It is very rare that the solution to a problem will present itself accidentally, although it does happen occasionally. Power quality work has its rewards. One that the author cherishes the most is the satisfaction of knowing that he has left clients happier than when he first met them.

Index

197